Trajan's Hollow

Trajan's Hollow

Joshua G. Stein

GIOVANNI BATTISTA PIRANESI, "TRAJAN'S COLUMN,"

COPPER PLATE ENGRAVING, 1758

Foreword: Reproducing Monuments

DAVID GISSEN

Joshua Stein's *Trajan's Hollow* joins the numerous reproductions of Trajan's Column that collectively shape the meaning of the original monument in Rome. It is a volatile and germane time to think about reproductions and monuments. In the past twenty years, discussions of "digital" reproductions both extended and transformed historic ideas regarding how mechanical reproductions provided ways to access, understand, and preserve monuments. The most recent discussions of reproductions emphasize how both mechanical and digital reproductions are slowly transforming the meaning of monuments—in often profound ways. This brief essay considers some of these insights and how Joshua's project relates to them.

One hundred and ten years ago the Viennese art historian Alois Riegl defined monuments as those enduring witnesses of time that become monuments due to their "age-value" or "artistic value." The former term describes artifacts that become admired, and ultimately preserved, by the simple fact that they represent a particular time, while the latter term describes artifacts that represent a particular form of artistry or point of view. The terms are not mutually exclusive. Whether admired for artistry or historical legibility, monuments, in Riegl's definition, endure in place, but their surroundings transform. Riegl's definition of the monument emerged from his work in documenting the historic monuments of the Austro-Hungarian Empire. It remains an early and useful definition of the monument with broad influence in subsequent preservation and heritage documents. One first identifies those works of historic or artistic value and then documents them through reproduction processes.

While Riegl's definition of the monument remains sensible, other qualities might be ascribed to monuments that upend relations between monuments and their documentation. Are monuments first recognized for their historical and artistic contributions and then reproduced so that these qualities can be admired? Or, as Jean Baudrillard, Bruno Latour, Thordis Arrhenius and Jorge Otero-Pailos suggest, are artifacts reproduced in such a way and in such quantities that they become monuments to be appreciated for artistic and historical value? The relationships between monumentality and reproducibility are volatile and non-linear.

The reproductions of Trajan's Column that Joshua Stein's project joins are extensive and include virtually every technique from the 18th to 21st centuries: 18th-century copper engraving, early 19th-century British and French plaster casts, English and American rubbings, Daguerreotypes, black and white photographs, color photographs, polyester casts, photogrammetric models, laser scanning, polychromatic reconstructions, and typographic reconstructions. These reproductions enhanced calls to preserve the Column and to canonize the work in the pantheon of antique monuments.

If monuments can be defined as historical artifacts in a constant state of reproduction, then another quality of monuments is their distribution. While Riegl could define monuments as singular artifacts of human creation in time and space, artifacts also become monuments when they appear everywhere. Reproductions of Trajan's Column exist in London, at the Victoria and Albert Museum, in Bucharest, at Romania's National History Museum, in a printing shop—R.R. Delaney—in Illinois, and in Rome, among other locations. I can write this essay in the all-caps typeface known as "Trajan" and that is based on Edward Catich's fastidious, mid-20th-century analysis of the inscription on Trajan's Column. The original inscription that includes almost every letter of the Roman alphabet, has itself been reproduced countless additional times. We can experience aspects of the inscription when typing in the fonts Baskerville, Bodoni, or Optima, as well.

This distributed quality of monuments also extends to their original stones. Many monuments have been moved far from their original locations and are beheld by new audiences—from the various Egyptian obelisks in Rome to the Dendur Temple in New York City's Metropolitan Museum of Art. Generally speaking, the distribution of monuments tends to move monuments from the world's poorer locales to its wealthier ones. Wealthy collectors and antiquarians cut and disassembled antique monuments into pieces, distributing them to many locations—the Parthenon Marbles in the British Museum being one of the most infamous.

When Napoleon occupied Rome, he ordered his ministry of culture to disassemble Trajan's Column so that the monument could be transported back to Paris to become part of renovations of La Place Vendôme. His plans were never carried out but two replicas of the Column were produced during the Napoleonic occupation: the plaster cast of the monument that now sits in the Victoria and Albert Museum and la Colonne Vendôme in la Place Vendôme, Paris. The latter is a bronze interpretation of Trajan's Column (versus a direct copy) and celebrates the victories of Napoleon over the Hapsburg Empire. The Parisian copy of Trajan's Column has been reproduced countless times, and the "original" sitting in Paris is actually a reconstruction. The original Colonne Vendôme (can we use that term?) was destroyed in a spectacular act of protest by La Commune de Paris in May, 1871 and rebuilt a few years later.

Why are monuments so reproduced and so distributed? One might think this is solely due to their aesthetic appeal. But Thordis Arrhenius defines monuments as inherently fragile artifacts always threatened with destruction, disassembly, and erasure. Again, this is a counter-intuitive argument, because most people think of monuments as enduring. In Arrhenius's formulation, reproduction becomes a form of preservation and ensures the continuation of the original—like distributing financial or resource risks into many different locations and spaces. If the Parthenon exists in five different cities then its destruction can be more easily averted. If the center of a city—like Warsaw—has been visualized in paintings and photographed hundreds and maybe thousands of times, then these images can be used to reconstruct it. And this is precisely what happened after it was destroyed by the German armies in World War II.

Contemporary theorists of the reproduction of monuments such as Bruno Latour, Adam Lowe, Brendan Cormier, and Mari Lending argue that these reproduction processes create an ontological crisis in the monument. They confuse a facile and earlier distinction between copies and originals. If someone such as Walter Benjamin could argue that mechanical reproduction was in a dialectic with originals and their "auras," post-modern theorists of monuments and reproductions argue that originals and copies have more equivalence.

Bruno Latour and Adam Lowe argue that reproductions confuse the relationship between originality and authenticity. A plaster cast made before a monument became badly weathered has more of the original surface features than the authentic monument. Neither the original or the copy gives the whole picture, and a scholar would likely want to view both to understand a monument. Post-post-modern theorists of the copy such as Mari Lending dispense with the distinction between originals and copies altogether and argue that copies and originals are each "productions." In my own writing on this topic,

I prefer the distinction between "models" (instead of original) and "translations" (in place of copies).

Joshua Stein's *Trajan's Hollow* pushes the boundary of what we might consider a reproduction, production, or translation in that virtually all of the reproductions of Trajan's Column cited in this essay resemble some experience that could be acquired of the monument in Rome. But I think its reproduction-like quality is one of its strongest characteristics and gives the project a documentary ethics. Joshua gained access to the interior as part of his Rome Prize fellowship and felt that his experience inside Trajan's Column was missing from virtually all reproductions of it. Thus, his project extends that experience to a larger audience. This is a very provocative idea of a reproduction—as a supplement to an original that exists, but that can never be experienced.

Reproductions often stage the disciplinarity of artifacts that have otherwise ambiguous disciplinary associations. For example, most reproductions of medieval altar pieces (works of painting, sculpture, and furniture) reproduce them as part of the history of painting. And Joshua's project continues this quality of reproductions by reproducing Trajan's Column in a 21st-century sense of architectural experience. This is also provocative as the original was a hybrid work of architecture, sculpture, and painting—created long before Western disciplinary separations existed. Most people forget that the Column was completely painted with bright polychromy, and the experience of the Column as a work of painting has been lost.

Reproducing Trajan's Column as a work of architecture presents a problem: if we stripped away all the sculptural surfaces of the outside and simply reproduced the interior and exterior as bare marble surfaces, we would be left with a monument solely appreciated for its historical versus artistic value. The monument's construction has more resonance with contemporary ideas about sculpture than architecture (ie. the enormous rings of stacked and carved stone). And the space of the interior and its light apertures are not particularly architecturally interesting or innovative, especially compared to other Roman buildings.

To get around these problems, Joshua introduces various fissures to make the connections between the inside and outside of the monument more complex, comparative, and intelligible. He takes something that is as much architectural as it is sculptural and painterly and reproduces it into an architectural disciplinary framework. His work forces us to appreciate this ambiguous and volatile monument as architecture.

Joshua's project makes me wonder if a painter might reproduce the monument solely based on 21st-century concepts of the discipline of painting. Could the monument be reproduced as a polychromatic surface, dispensing with any sense of sculpture or architecture? Could a sculptor reproduce the monument as figures without any sense of paint, support structure, or space? Typographers have certainly done a good job of reproducing the monument solely as a work of typography.

As interesting as these disciplinary thought-exercises might be, it could be equally provocative to see a reproduction of the monument as a place to stage alternative forms of interdisciplinarity. Roman and Greek societies did not easily distinguish art, architecture, and sculpture in the way we do, so why is it necessary to project disciplinary values onto these works? What is gained?

Nevertheless, and as this essay points out, one of the many ways monuments become monuments is through their reproduction and distribution. Joshua's project joins a very large corpus of reproductions of Trajan's Column that in their totality create the aesthetic experience of this monument.

Most people experience canonical works of art, painting, sculpture, and architecture through reproduction. Thus, reproductions are not only important due to their ubiquity, but also because most people experience the values of the person or people who made them. In *Trajan's Hollow* we experience Joshua's generosity when he reproduces a part of a monument that most of us will never be able to experience in Rome. This idea of generosity has so much potential, and I hope Joshua continues to explore it and to make us see it as an integral quality in reproductions of all forms of culture. In turn, and as the above writers demonstrate, this will transform how we think about monuments.

VIEW FROM THE BALCONY OF TRAJAN'S COLUMN (TOP) AND LOOKING FROM
THE BALCONY BACK DOWN THE COLUMN'S SPIRAL STAIR (BOTTOM)

Introduction

JOSHUA G. STEIN

This book is the result of seven years spent processing a rare twenty-five-minute visit to a historical monument constructed nearly 2,000 years ago, from materials formed twenty-five million years before that. What can be learned today from a storied Roman monument that has been obsessively documented over the course of two millennia? Trajan's Column, one of the great masterworks of Roman antiquity, has been examined scrupulously by historians and archaeologists, and admired by tourists and Romans alike. Despite its prominence within academic discourse and its monumental stature within the city, few would use the word "architectural" to describe the Column; few even realize the monument is habitable. But I was to discover that this hollow cylinder, with its internal spiral staircase, shapes a nuanced interior experience, unknown (or unappreciated) by most. *Trajan's Hollow* uncovers aspects of the Column that have been neglected amidst all the historiographical attention focused on the monument. In so doing, it reveals the lacunae in many of the dominant paradigms of historical inquiry. By temporarily turning away from the abstract geometries, proportional exercises, and visual outlines that so often define our study of the ancient world, this research proposes to map the surfaces and mine the interiors of these monuments, revealing patterns and particularities that might otherwise be written off as anomalies or mistakes. In a contemporary culture increasingly inundated with knock-offs, contested samples, remixed and rehashed styles and forms, this project seeks to establish a methodology for the transformational use of our inherited patrimony through playful yet rigorous documentation, modification, and recontextualization.

Trajan's Hollow has one straightforward objective: to physically reconstitute all aspects of Trajan's Column that have been hidden, forgotten, or excised from public view, collective memory, or public record. This simple goal becomes complicated by the thorny challenge of defining exactly what constitutes Trajan's Column in the first place. Next follows the task of identifying the best method to document that which is missing. As an object of study, this monument was selected precisely because of the many complexities it reveals, thus demanding more innovative techniques of exploration. It was clear from the outset that to properly examine the role of materials in the Column, tangible production would constitute an essential aspect of the process. In addition to documenting the analysis of Trajan's Column and its reconstructions *Trajan's Hollow 1:10* and *Trajan's Hollow 1.5*, this book offers a platform for a set of architectural mini-manifestos, with the physical artifacts operating as a test bed for translating these proclamations into action.

My encounter with Trajan's Column was a fundamentally indirect one in that I discovered this monument without any real interest in the monumental. Instead, I was led to the Column by tracking the mineral veins that saturate our geological and urban strata and their emergence into architectural form. I began my Rome Prize research scouring the Eternal City for plaster: plaster replicas, ornamental plaster details, and gipsotecas. Although I do have a love for the visceral qualities of this changeable substance, it was instead what plaster represents for architecture and the discipline's conflicted relationship with history that captivated me. Before leaving for Rome, I began an archaeology of the reproduction of sculptural form and ornament, especially the plaster casting systems developed from the classical period through the early 20th-century, and an examination of the potential these techniques hold for contemporary design. The relatively fragile nature of plaster meant that I was moving through Rome searching for second-hand accounts of antiquity instead of the "real" thing. I was therefore more interested in architecture's documentation of the past than any authority history might hold over the present. Plaster led me to the Museum of Roman Civilization, an institution composed almost entirely of fakes and surrogates: plaster casts, fiberglass casts, reconstructions, and miniatures. Here I discovered the 1861 plaster casts of Trajan's Column, arranged horizontally (a radical shift from the Column's emphatic verticality) in a long corridor connecting two wings of the museum. I would return to the galleries several times, slowly realizing that the subject of my study would be the Column and its long history of copies, and ultimately, that the outcome would be adding to this catalog through my own reconstructions of the Column.

Plaster's liquid phase offers a method of documenting the past that is not dependent on human interpretation—or at least is less so than sketching, diagramming, or writing. It is more akin to the digital scanning techniques that are now used to document antiquities and archaeological finds. This analogous relationship between scanning and casting allows this historical research into plaster casting to open potential explorations applicable to today's emerging technologies. Plaster is perfectly suited to the precise indexing of information, yet it does so incompletely—and impermanently. This process establishes a paradigm where history is not simply preserved, but is materially recycled and redistributed. In the analog processes of plaster casting, the necessity of contact with the original object ensures a replication of material information, capturing texture easily but form with more difficulty. This study of transference has broad implications for architecture, a discipline that is continually engaged in an act of translating or reproducing form—from the scale of the model to full scale, from virtual to physical, from abstract to tangibly specific form, and vice versa.

Through the Column's casts, I then found my way to the Column itself and discovered a surprisingly incomplete perception in the public and academic consciousness. Trajan's Column is well known within Rome, and yet even many Romans are still shocked to hear that the Column is, in fact, habitable. Of course the Column and its grounds are off-limits to the public, so it is only anecdotally that one would hear of the spiral staircase that brings the lucky visitor to the vertiginous overlook at the top. Through the American Academy in Rome, I was able to submit a special request to visit the Column. I was never granted official access to visit and document the interior, but I was able to join a maintenance visit by a local utility company. There was no duration designated for this visit until time had abruptly run out. Realizing that this would most likely be the last opportunity to visit the inside of the monument, I quickly documented the interior as best as possible with help from my Academy colleagues. After exceeding the patience of our guide, I was pushed down the last few turns of the spiral stairs and brusquely escorted off the property.

But the visit afforded an incredibly rare opportunity to experience what most people do not know exists. The interior is cool and pleasant on a hot Roman day. The constraints of the staircase are tight, and yet the embrasure of the windows creates a more generous set of wedge-like spaces than one might expect based on the miniature, barely-perceptible windows encircling the column exterior. The chiaroscuro play of sunlight from these apertures accentuates the chisel marks of the ancient masons. There are also other signs of human occupation over the last two millennia: graffiti from past centuries, crude conduits for lighting, and the preservation experiments of contemporary conservationists. As virtually no documentation of the interior exists, this quick expedition into the column may provide the first indications to the public of the hidden experience within this well-known monument. But how could it be that no one had ever before properly detailed the interior of the Column? The extreme imbalance between the amount of information of the Column's exterior versus its interior points to a paradigm for documenting classical monuments that prioritizes an essentialist visual mode, manifested either through plan analysis or perspectival sketching.

Trajan's Hollow attempts to backfill the gap created by this dominant tradition of architectural historical analysis—in particular the study of ancient and Renaissance architecture preoccupied either with identifying overarching tendencies through comparative analysis, or with the pictorial representation of ruin and degradation. The agenda of my inquiry is not to undermine these traditions but rather to complement them by resuscitating an undercurrent of more embodied perspectives towards history. One easy foil might be Rudolf Wittkower and the legacy of diagrammatic analyses that follow him, from Colin Rowe to Peter Eisenman. This tradition places attention on plans and elevations (sometimes perhaps even a section or axonometric view) redrawn to reveal the underlying geometric relationships exhibited (or intended to exhibit) by Palladio. This paradigm then retroactively expands backwards in time to encompass a wide range of architectural works from different epochs, but especially classical and Renaissance, mannerist and baroque (it is much rarer to find similar geometric analyses of Gothic plans or facades). These exercises promote an understanding of endlessly reconfigurable formal permutations—a powerful paradigm, yet one completely dislodged from the realm of matter.

The opposite extreme perspective is that of the picturesque—obsessed with the degeneration of classical formal purity. While this tradition is concerned with transformation over time, its fascination ultimately still continues a primarily visual obsession with architectural form. It is also a tradition that is fundamentally exterior, emerging from landscape design theories that offer up architecture as an instigator for procession through a visually calibrated and manicured environment. In this scenario, architecture is less desirable for its habitable qualities than its role as folly or foil for landscape. Rarely do the studies of ancient ruins attempt to understand the precise geometric implications of weathering beyond fetishizing the aesthetics of decay. And while there are certainly technical studies by conservationists that seek to understand internal material failings, these studies tend to reinforce the assumption that there is some ideal or original condition. *Trajan's Hollow* pushes against the oppression of the visual both as geometric and pictorial—either the tyranny of the god's-eye view obsessed with abstract patterning behind form, or that of the perspectival eye in the landscape concerned with exterior aesthetics. Of course this visual regime has proven useful—the tactile, the haptic, the scan seem less productive as tools to sum up the essence of an object than the visual, either diagrammatic or pictorial.

Trajan's Hollow attempts to penetrate beyond the superficial image without falling prey to the abstraction of the diagrammatic. Operating as a core sample of both material properties and human haptic experience, this process immediately delves into the Column, unable to remain aloft or afar. While so many studies of historical artifacts seek to read the object through patterns in the oeuvre of the architect or the historical movement, this study instead draws corporeal and experiential connections out of the monument itself. Instead of analyzing the historical monument beneath a superimposed filter of geometry, teasing out an idealized and immaterial architectural schema through interpolation between instances of other similar projects by the same architect or of the same period, *Trajan's Hollow* uses a historical artifact as a source from which analysis might extrapolate an interior realm of seemingly unclassical concerns. This blind mapping connects to geology, craft technique, human wear, and mineral transformation: the figural narrative of the Column becomes connected to its chemical and meteorological contexts as much as it is to Roman or Greek sculptural traditions. Relieved of the obligation to identify connections to other projects, past or present, within the architectural canon, this investigation's seeming myopia begins to reveal subterranean connections to ecologies (human and non) that push at the boundaries of architectural discourse. This inward inquiry that ultimately seeps outward—a centripetal focus that translates into centrifugal expansion—can be described as a study of interiority.

This concept of interiority reconciles the two seemingly disparate notions of phenomenological embodiment and ecological embodiment. The first indexes the anthropocentric experience of the architectural subject, the sensing human body as it moves through and perceives space. The simple act of examining Trajan's Column as a habitable work of architecture rather than a monument privileges its experiential and atmospheric qualities, more attuned to haptic understanding than a visual reading. The inside of the Column represents and embodies all that is easily overlooked. It is phenomenology's embodied subject that experiences the visceral vertiginous winding upward through the Column, while remaining sensorially grounded in the earth via the massive construction of excavated marble drums.

The phenomenological conception of connectivity finds support in a second notion of embodiment, which is decidedly non-anthropocentric. Here, the material body could be understood in the manner that ceramists speak of "clay body," simultaneously integrating aesthetic characteristics, behavior, chemical composition, and provenance. The interiority explored by *Trajan's Hollow* also embraces the model of "embodied energy," the tally of all of the operations (usually industrialized) that have been enacted upon a discrete building component and the energy outputs associated with these processes. If one includes "operational energy," incorporating all future energy expenditures for upkeep, this mindset not only traces backwards but projects forwards in time. This paradigm denies the assumption that one might understand an object, commodity, component, or work of architecture as discrete from its spatiotemporal context, and renders irrelevant the intentionality of the architect. How might an embodied understanding of form manifest itself within a dynamic ecology of forces, both natural and human? In the case of Trajan's Column, we first recognize that there is no such thing as an ideal moment in time from which to ascertain the "best" condition of form. It is therefore equally relevant to examine the Column's Carrara marble at the moment it is buried under fifty feet of "overburden" in the Apuan Alps or while it's eroding from the exterior of the Column, slowly cascading down its stairs and lining the storm drains of Rome, as it is at the moment the monument was opened to a select public in 113 CE. While the aesthetics of the ruin assumes a set start date after which weathering and degradation set in, my study is interested in durational transformations that are continual, with phases of different intensities, but without an idealized point of origin. The overlooked inside of the monument offers an invitation to examine its interiority as both embodied human experience and a materially informed artifact with the inherent qualities, journey, and transformation of the matter that constitutes the Column and its progeny.

At the outset, Piranesi's uneven consideration of the Column is also set up as a foil. At moments, his documentation of the exterior seems as intricate and impartial as the digital input devices employed for similar purposes today. His elevational drawing of the Column's pedestal (see pages 293-295) operates as a scan, including all of the blemishes of time without fetishizing the Column's ruin. But Piranesi's detached precision is inconsistent and most importantly does not extend to the interior, where he instead excises specificity and actuality in favor of producing a generalized and rationalized speculation on the assembly techniques employed in the construction of the Column. However, his section drawings do dramatically, if not precisely, reveal the existence of the Column's interior and its connections to the exterior through its forty apertures. Ultimately, Piranesi becomes more a coconspirator in this study than an oppositional nemesis. While his obsessive studies of ancient Rome leave open a window for more study on the Column, his creative interpretation—his intentional misreading of the city and its monuments—also serves as a model project, irreverent and transformative.

The intensely physical and interventionist strategy of this analysis cannot be supported by diagramming and drafting alone. Instead the tools of inquiry are an arsenal of chisels, pantographs, digital scanning and modeling software, buckets and molds for plaster casting, and 3D printers. The research is not only a study of Trajan's Column but its reproduction, culminating in two critical facsimiles of the historic Column. The first is a digitally fabricated, 1:10 miniature of one of the Column's drums that, through its production, retraces the material journey of the original Column's marble quarried in Carrara, while also manifesting the traces left by current digital machining techniques. The second is a cross-section of the Column produced as a six-meter diameter plaster bas-relief floor cast installed at the American Academy in Rome and constructed at 1.5 times the scale of the original. This habitable installation renders Trajan's Column legible as architectural space while transposing its renowned decorative frieze into phenomenologically apprehensible terms.

Trajan's Hollow crystallizes three related research pursuits: the tectonics of poché, the role of subversive reconstruction, and the use of scalar operations within architecture. In the face of architecture's recent obsessions with novel building envelopes created from synthetic and engineered elements, it is an opportune moment to instead reconsider the potential of thickness and physical weight, as well as the metaphorical weight of history and tradition, as a productive opposition within contemporary architectural discourse. This material resistance entails recovering the value of opacity and obscurity in an era of diagrammatic and disembodied transparency. It proposes an architecture that reckons with the cultural, historical, and political associations of tradition, and that questions the technological imperative of progress. Instead it directly engages the traditions of craft (stonemasonry, quarrying, plaster casting) and their potential to once again connect architecture to its ecological context. The essay "Pocket Landscapes: Trajan's Monument to

Poché" examines the Column's spiraling sequence of partially obscured openings to its physical and cultural context, which has transformed dramatically over time. This rhythm of incomplete revelation offers an alternative to the all-knowing bird's-eye view of history and allows the Column to embody multiple readings that do not necessarily synthesize into a neat, comprehensive whole.

Trajan's Hollow also provokes a discussion of the techniques of imitation and reconstruction, addressing the increasing importance of critically analyzing such practices in an age in which form and texture can be instantly captured and reproduced digitally. As architectural discourse swings between advocating for the novel production of form and ready-made reference to the past, this research aims to develop an alternative model in which the increasing complexity of technology offers, and in fact demands, output that is both interpretive and generative. The essay "The Wayward Cast: Gipsotecas, Digital Imprints, and the Productive Lapse of Fidelity" plays on Georges Didi-Huberman's distinction between the *imitation* and the *imprint* to open a critical discussion of reconstructions in light of emerging technologies. This inquiry is timely in the context of contemporary debates over historical reproductions and the fate of the existing cast collections housed in various international museums. This discussion took on a new significance with the advent of the digital scan and reproduction of the Palmyra Arch destroyed by ISIS, and through exhibitions such as the V&A's *A World of Fragile Parts*, featured in the 2016 Venice Architecture Biennale. The relocation and rehabilitation of the Metropolitan Museum of Art's plaster cast collection, restored after years of neglect and only recently finding a home at the Institute of Classical Architecture & Art in Manhattan, also points to a contemporary interest in reproductions that have been, until recently, considered outmoded and irrelevant.

The two creative reproductions, executed at scales both smaller and larger than the original, and the accompanying essay, "The Agency of Scale: Employing Scalar Shift as a Design Strategy," stake a claim for scale shift as one of the primary tools of the architectural discipline. While rarely identified as such, scale shift is an operation that offers great potential for the architect. In addition to the architect's necessary ability to connect information at multiple scales, the intentional removal or introduction of information required in shifting scales offers the opportunity to expose, develop, and encourage new allegiances and associations, effectively "reprogramming" the original design. *Trajan's Hollow* offers a method of carefully oscillating between the a-scalar space of virtual modeling and the precise and singular scale dictated by material operations and craft techniques.

This book is composed of two major sections, "Part I - Monuments" and "Part II - Reconstructions." The first half examines the Column through somewhat standard methods of documentation while the second operates as an analysis through production. While the book does not necessarily need to be read sequentially, the earlier chapters provide a context for design decisions in the second section that might otherwise seem idiosyncratic. The first chapters attempt to unravel the complex history and actuality of the Column, including its progeny of copies and reproductions. Michael J. Waters's essay "On Architectural Materiality: Between Trajan's Column and *Trajan's Hollow*" situates the Column as part of a tradition of embodied material memory. Scholars and aficionados of Rome and Roman architecture may be particularly interested in the new documentation of the Column's interior provided in "Hidden Trajan." The second section documents *Trajan's Hollow*, the creative reconstruction of the monument described in the first section. It is every bit as precise as the first half, but the analysis is embodied through physical techniques, rather than drawings or photographic images. Michael Swaine's "A Set of Directions for the Reader: 185 Steps" attempts to textually reconstruct the experience of climbing the Column's interior and all the associations that the space provokes. The two physical reconstructions of the Column attempt to do the same through material means.

Although historical case study has a well-worn trajectory in architectural studies, this research proposes employing new technologies to stage a fine-grained examination of history. Rather than compiling a series of historical examples to define a type (a worthwhile, yet welltrodden exercise), *Trajan's Hollow* provides an unironic, in-depth study and re-creation of a singular monument which triggers a cascading series of revisions to our assumptions in reading artifacts, shifting from the visual regime to material actuality. The research directly tackles architecture's fascination with historical imitation (most typically produced in this case through quotations of the Column's iconic silhouette) and proposes a material rather than merely visual transference of historical information. By displacing our desire for fidelity from the reproduction of iconic profiles to the reanimation of historical materials worked by human hands and ecological forces, this paradigm shift offers the potential not only to produce novel form, but to reprogram inherited patrimony, ultimately dissolving the disciplinary dichotomy between acritical form-making and revisionist postmodernism.

Part I—Monuments

Trajan's Column

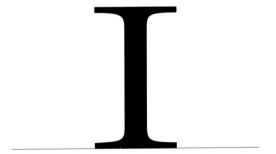

COLUMN

19 drums in Column weighing 32 tons each
Monument constructed of 29 blocks of Luna marble
185 steps
40 embrasured windows in Column
3 embrasured windows in pedestal
Column diameter at bottom = 3.70m
Column diameter at top = 3.20m
Entasis begins 1/3 of the way up (at 9.216m)

NARRATIVE RELIEF

Spirals counterclockwise around shaft 23 times
Total length = approximately 200m
Scroll increases in height from bottom (0.89m) to top (1.25m)
Figures increase in height from bottom (0.60m) to top (0.80m)
155 scenes
Trajan appears 59 times

*Piranesi both "scans"
and reconstructs
the exterior of the
Column from top
to bottom through
a series of intricately
detailed engravings.*

GIOVANNI BATTISTA PIRANESI, "VIEW OF THE FRONT OF TRAJAN'S COLUMN," PLATE III,

FROM *THE TROPHY OR MAGNIFICENT SPIRAL COLUMN*, 1774–75, COPPER PLATE ENGRAVING

(CONTINUED ON RECTOS OF PAGES 29–37)

Piranesi's analytical axonometric and sectional drawings are more concerned with conjecture on assembly techniques than material accuracy.

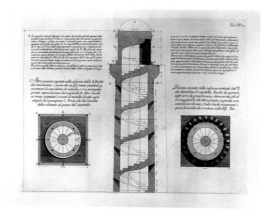

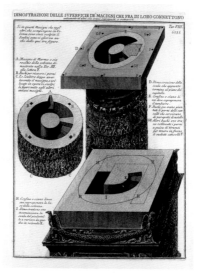

G.B. PIRANESI, "SECTION OF TRAJAN'S COLUMN," PLATE IV (TOP), AND
"ILLUSTRATION OF THE DRUMS AND HOW THEY WERE CONNECTED,"
PLATE VIII (BOTTOM), 1774–75

Piranesi's elevational drawings, however, reveal a sporadic interest in depicting the existing condition of the Column before him, blemishes and all.

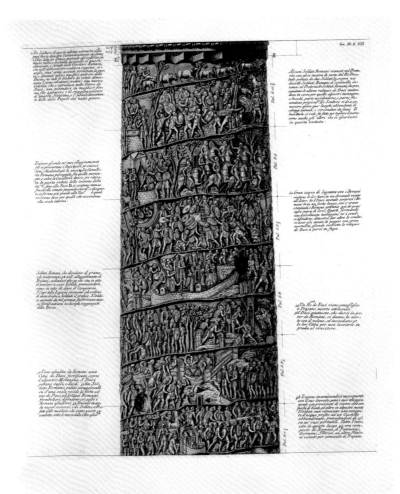

G.B. PIRANESI, "VIEW OF THE FRONT OF TRAJAN'S COLUMN," PLATE III, 1774–75

Piranesi documents the Column from two distinct perspectives: first, according to its visual appearance at the moment that he produces his engravings in 1774 based on empirical observation, and second, following his speculation concerning the architect Appolodorus's methods of construction employed over 1600 years earlier.

On the one hand, he shows the Column as intricate sculptural object. His highly detailed renderings of the exterior capture every aspect of the sculptural relief, including robber holes or the damage sustained during the pillaging for materials during the Middle Ages.

On the other hand, he demonstrates the Column as a feat of construction technology. In these drawings, Piranesi edits out existing specifics of the Column in favor of projecting a coherent intention on the part of Apollodorus—the complexities of material, technique, and time whitewashed by the idealized geometries of the ancients.

Yet the Column is much more than a vehicle for sculpture. The simple formula of an honorific column masks a building—for that is what it is—of some complexity, incorporating an entrance door, a vestibule, a chamber, a stair, windows and a balcony in the form of a capital. Indeed, as a work of architecture in its own right, it has been imitated widely.[1]

Mark Wilson Jones

The gap between these two different modes of documentation elides the interior experience that renders this Column more that sculpture or monument but architecture.

Between the exterior visual and the idealized analysis of geometry exists the actuality of human experience and material qualities—
Trajan's Hollow.

1 *Principles of Roman Architecture.* (New Haven: Yale University Press, 2000), 161.

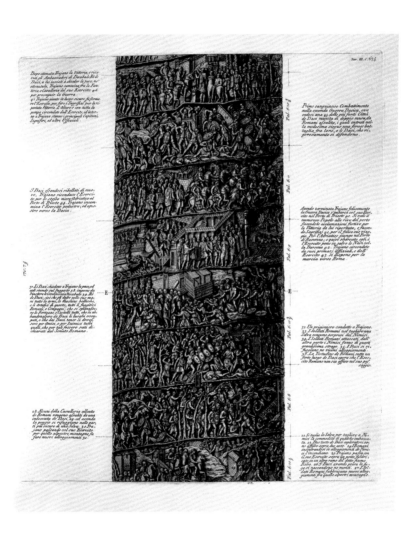

G.B. PIRANESI, "VIEW OF THE FRONT OF TRAJAN'S COLUMN," PLATE III, 1774–75

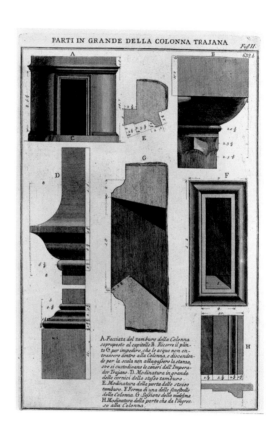

G.B. PIRANESI, "VIEW OF THE FRONT OF TRAJAN'S COLUMN," PLATE III, 1774–75

Piranesi's elevations of the pedestal produce an even more accurate "scan" than the top-to-bottom elevation, revealing a meticulous study of the Column's material actuality.

G.B. PIRANESI, "PEDESTAL, AND BASE OF TRAJAN'S COLUMN," PLATE IX, 1774–75

While the top-to-bottom elevation includes indications of degradation, it also reconstructs the decorative bas-relief, idealizing both the grandeur and ruin of the Column.

Veduta del prospetto principale della Colonna Trajana
A. *Piano moderno.* B. *Linea anticamente indicata sopra lo stesso sasso, che dimostra dov'era il piano antico della Colonna*
Le Linee de punti E indicano l'altezza delle pietre che compongono la Colonna

G.B. PIRANESI, "VIEW OF THE FRONT OF TRAJAN'S COLUMN," PLATE III, 1774–75

TRAJAN'S COLUMN VIEWED FROM BELOW

On Architectural Materiality:
Between Trajan's Column and Trajan's Hollow

MICHAEL J. WATERS

There has been a recent obsession with materiality in architecture. Why have architects begun to think once again about the potency of matter as a bearer of meaning? After all, descriptions of buildings have for over 2,000 years focused primarily on materials. There is no singular answer, but it is certainly part of a broader societal trend, one that may be partially a reactionary movement to the Digital Revolution, just as the Arts and Crafts movement was to the Industrial Revolution, or even an outcome of the rising problem of ecological scarcity. Yet within the architectural discourse, many have suggested it is a response to the removal of materials from the conceptual process of architectural signification with the rise of Modernism. While early 20th-century architects celebrated the arrival of an array of new materials, many of which they claimed dictated form, architecture increasingly moved toward abstraction. At the same time architects fetishized individual materials (glass, steel, concrete, and even marble), many moved towards an architecture in which meaning was located in the abstract, geometrically pure design. Yet the marginalization of materiality in architecture is by no means purely the consequence of Modernist dogma or Deconstructivist manifestos. It was in the Renaissance that the humanist Leon Battista Alberti first theorized an architecture fully formed and perfected in the mind, which was only later realized in and corrupted by physical materials. In his conception of design, form and materials were conceived of separately and independently. While this hylomorphic dichotomy was supported by Aristotelian natural philosophy, in terms of architectural practice, the concept that materials were irrelevant to design and subservient to form was completely new with little basis in contemporary practice. The separation of architecture from its inherent material essence thus has its origin in the 15th-century. That said, it has been only more recently that this concept has begun to significantly shape architectural production.

This divide between form and material did not exist in antiquity. Design in the ancient world, while grounded in theoretical concerns and architectural norms, was both a physical and mental activity. It moreover was directly linked to the material process of construction.

Scale-drawings (plans, sections, and elevations) were used only occasionally. Information was instead largely conveyed to builders through text, models, templates, on-site drawing, and, most of all, oral communication. The architect, therefore, often shaped a building as it was being erected. The case of the Column of Trajan was likely no different. The enormous monument must have followed a general design that stipulated its overall form: a large honorific column approximately 100 feet tall and 200 digits wide set on a plinth with a corkscrew staircase inside and a spiraling relief outside. Some details such as the number of stairs per revolution would have been decided before construction began. Yet many elements of the Column were only figured out during the process of construction, including the internal transitions of the stair, the number of blocks used in the plinth, and the height of each column drum. We have been conditioned by modern building practice to see features such as these as changes to the initial design or derivations from the ideal. This obscures the fact that design was a physical activity that continued during construction. But more than that, by focusing on the question of design, we have lost sight of the reality that much of the monument's original meaning was imbedded not in its abstract form but rather in its materiality, that is, its corporeal essence as shaped by its physical materials, the nature of those materials, and the processes and techniques essential to its creation.

The Column of Trajan is built of twenty-nine blocks of Luna marble which weigh together over a thousand tons. The pedestal is composed of eight blocks atop which stands a column base, a shaft built of seventeen drums, a capital, and a crowning cap assembled from two pieces of stone. The facture of the monument was in many ways dictated by the choice of material. As Mark Wilson Jones has noted, one would expect the plinth to be monolithic, like the later Column of Marcus Aurelius. Yet a single block of marble this size would have weighed nearly 100 tons. Quarrying, transporting, and lifting this large a monolith was beyond the limits of Roman engineering capabilities. It was this restriction that made a monolithic pedestal impossible. Yet despite this material constraint, the same builders decided to construct the Column out of nineteen drums, each roughly five feet high, rather than hundreds of smaller pieces of marble. This choice may have been motivated by structural concerns, but it is also indicative of a preference for monolithic facture. Romans during the Imperial Period loved monoliths. While during the reign of Hadrian sixty-foot marble monolithic column shafts were brought to Rome, producing a monolithic Column of Trajan could never have

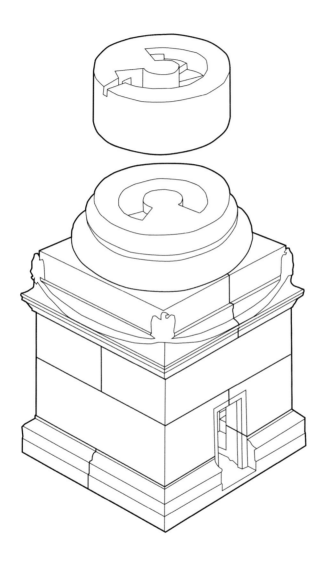

MARBLE BLOCK FACTURE OF TRAJAN'S COLUMN

been in the realm of possibility. Nevertheless, rather than building the Column out of many small blocks of stone, the builders used as few pieces as possible imitating the Greek method of fashioning marble columns out of drums, but at a previously unimaginable scale. Moreover, the razor thin joints between each drum are almost completely obscured by the spiraling sculpted frieze. From a distance, the Column appears monolithic. It is only on the inside that these seams are clearly visible.

The choice to build with monolithic drums necessitated quarrying and transporting enormous cylindrical blocks of Luna marble weighing between thirty and fifty tons from the upper Apennines to Rome. Each massive drum was placed on ships and sailed down the Tyrrhenian coast to Portus. There they were transferred to barges, shipped up the Tiber River, and eventually carted to the Forum of Trajan. The builders then began the process of construction by carving out the window embrasures, stairs, and ceiling soffit from each monolithic drum. In excavating these architectural elements, the stonemasons reversed the normal method of building by first working in a subtractive mode hollowing out the interior of the structure before it was erected. This exceptional procedure made the Column more like a work of sculpture than architecture. Yet unlike a rock-cut tomb or temple, once carved, each drum had to be lifted in place and perfectly aligned with the one below it. As Lynne Lancaster has examined in detail, this process of erection was a herculean task unlike any seen in Rome before. Typical Roman cranes could not lift blocks of this size. Instead, it is likely that an unprecedented four-masted lifting tower was built employing techniques Apollodorus of Damascus had used in siege machinery. This massive lifting apparatus must have measured over 130 feet high and featured colossal wooden trusses and a complex system of ropes and pulleys that enabled workmen to carefully lift, position, and lower each drum.

Enormous ephemeral structures such as this were often as impressive as the monuments they made possible. While scholars often overlook scaffolding, formwork, centering, and other similar temporary constructions, they were essential to the production of architecture. In many cases, they even dictated its form. The geometrical constraints of flat timber formwork, for example, established the shape of ancient concrete domes. Thus, when considering the materiality of a monument such as the Column of Trajan, we must understand not only its permanent materials but also the ephemeral structures that made its construction possible. These wooden structures were as much a wonder as the still standing Column.

Once the lifting apparatus and scaffolding was removed, it would not have been soon forgotten. The memory of the process of construction and the temporary structures that made it possible in this way imbued the final product with additional meaning. Visibility was not essential for signification. In fact, Martin Beckmann has suggested that Romans would have been more captivated by the extraordinary internal spiral staircase, which few would have experienced, than the exterior spiraling frieze, which was certainly visible if not completely legible. The sculpture of the Column, which has dominated nearly every study of the monument, was in this way subordinate to the hidden internal architecture. The meaning imbedded in the invisible is most clearly evoked in the dedicatory inscription found above the entrance to the Column. Commemorating its construction, the prominent inscription ends by proclaiming that it had been built "in order to show how lofty had been the mountain—and the site for such mighty works was nothing less—which had been cleared away." This opaque passage refers to the slope of the Quirinal Hill that was excavated for the construction of the adjacent Markets of Trajan. The emperor directly linked his Column to the incredible earth moving effort that made possible his larger Imperial Forum. This link between the removal of earth and the construction of the marble Column would have been plainly evident to any who lived in Rome in the early second century CE. Both were seemingly miraculous material achievements and testaments to the glory of Trajan and the power of Rome. At the same time, this connection was ephemeral. Once knowledge of these processes was forgotten, the Column of Trajan increasingly became about the final product. Yet it is clear from the Column's inscription that Trajan wanted to preserve the memory of this transient engineering achievement. The inscription therefore demands that we view this column as an impressive material monument created through an extraordinary process of construction. It was a process that involved not only leveling a hill, but also quarrying a mountain, shaping enormous blocks of marble, lifting column drums to incredible heights, and carving a continuous frieze. The true marvel of the Column was thus these staggering feats, all of which shaped its materiality for contemporaries.

Trajan's Hollow, visible at the American Academy in Rome for only a few weeks in the summer of 2011, at first seems to have little in common with the Column of Trajan. It has none of the same dimensions. It is not built of the same material, constructed in the same manner, or even decorated with the same sculpture. Nor could it ever be mistaken

DEDICATORY INSCRIPTION ON TRAJAN'S COLUMN

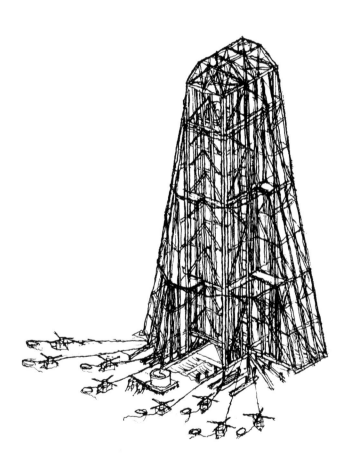

CUTAWAY RECONSTRUCTION OF LIFTING TOWER USED TO ERECT TRAJAN'S COLUMN

for a column. Rather, it is a work cast in plaster on the floor of a spacious studio that appears like an erratically routed elliptical ring attached to a central circle by a wedge-shaped appendage. In its form, material, and facture, it bears little resemblance to the monument it claims to be modeled on. Nevertheless, almost every element in the work is derived from the Column of Trajan. The project actually began as a digital copy of this monument, which was first modified through a series of virtual transformations. The figures in the frieze, for example, were translated into a series of slices, hollowed out from the exterior and interior of the outer shell of the monument. These computer-generated modifications were purely formal. Yet when this digital model was translated into material reality, something began to change.

Conceptually, the transition from digital to analog was made possible by taking a horizontal section of the computer-based model. This section was then turned into a five-centimeter thick plaster cast. But to reduce this procedure to these simplistic terms would be like claiming the Column of Trajan was merely carved of marble. Choosing to create a physical translation in plaster required a series of material gestures that transformed the project. First, a mold had to be created by projecting parts of the section onto a wall, then tracing these outlines onto pieces of polystyrene foam. These were then cut out using a handheld hot wire. This rather repetitive procedure in fact introduced the first element of variability into the final product. By cutting the foam in this manner, the edges of the mold were given a constantly irregular shape. When reversed in cast plaster, these ridges become a landscape of incredibly varied fjords anchored to the ground. The material of the formwork hence dictated the project's final form. While the precision of the digital original remains partially visible in the polished upper surface, the ad hoc casting process introduced material irregularity that obscures the clarity of its model. This messy process of creation was also made visually explicit through the preservation of the protruding unfinished joints. These vestigial fins, like the undulating surface of each peninsula, have little to do with Trajan's Column. Rather they are indexical material artifacts of the act of making.

Yet at the same time Trajan's Hollow was moving away from its model formally through the activity of casting, it was beginning to echo Trajan's Column more fully in its materiality. Like the ancient Column, its design was constantly changing through the process of construction that was dictated by material limitations. Just as the size of the Column drums of the ancient monument were limited by available lifting technology, the height of *Trajan's Hollow* was mandated by the thickness of commercially available polystyrene foam. The material of each project also introduced further variability. Despite the fine granularity of Luna marble, it has little tensile strength and becomes brittle when carved especially thin. The builders of Trajan's Column confronted this problem in each cylindrical drum. Specifically, each section has a horizontal joint that runs through not only the outer shell and inner core of the Column, but also the spiral staircase. This means that as the ceiling soffit reaches that transition point it reduces in thickness, in theory, eventually to nothing. In practice, this proved problematic, and in nearly every case, the marble broke at this juncture (see pages 66-67). *Trajan's Hollow* faced similar material difficulties. Plaster, like marble, has a thickness. When cast, it is prone to imperfections, such as holes produced by bubbles. These constraints directly shaped *Trajan's Hollow*. At one particular point, the initial design called for a sliver-sized projection. While this element, like the others, fans outward as it approaches the ground, it reduces at its top to such a degree that the piece became structurally deficient and broke off (see page 267). Although it was later repaired with wet plaster, the upper gouge still stands like the broken stairs of the Column of Trajan as evidence of the power of materials to dictate form. The ephemeral *Trajan's Hollow*, like the seemingly eternal Trajan's Column, thus derived a large part of its meaning from its materiality. For those who saw or even participated in its arduous production, the project was defined not by its theoretical underpinning or original digital model, but by its physicality and the material processes of its creation. While it did not require giant lifting machines or massive scaffolding, it was nonetheless a monumental undertaking. Just as with its ancient predecessor, much of that enterprise was never recorded; nevertheless, traces of it remained in the final product, visible to the careful observer. For those who never saw it in Rome, it will live on through this publication, and as such, move a step away from the physical world. But at the same time, like the colossal inscription that adorns Trajan's Column, this book will also serve as a perpetual reminder that to discover the essence of this work of architecture we must look beyond the surface and attempt to reconstruct in our minds the multifaceted nature of its now vanished physical materiality.

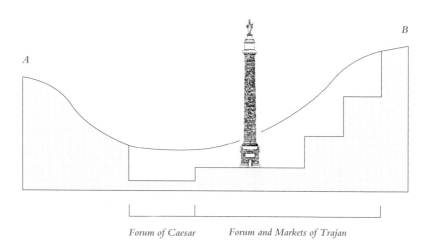

A

B

Forum of Caesar   Forum and Markets of Trajan

SECTION THROUGH TRAJAN'S FORUM SHOWING VOLUME OF EARTH REMOVED

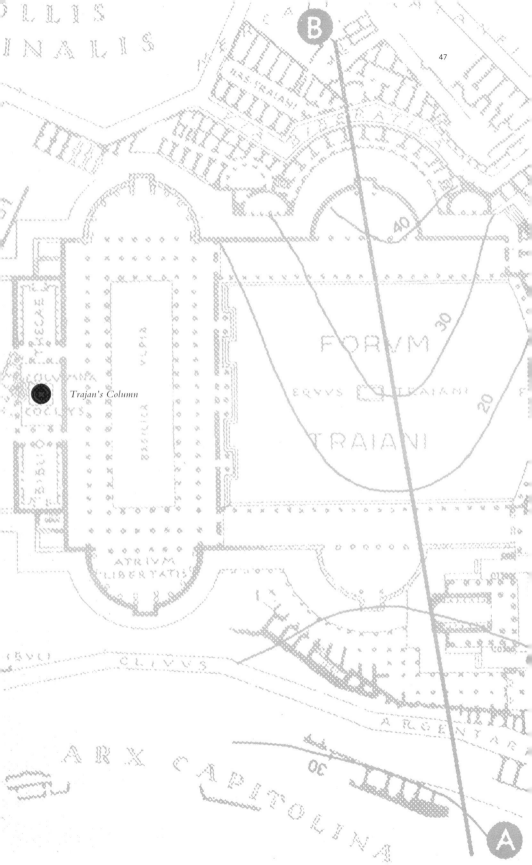

Trajan's Column

INTERSECTION OF NARRATIVE FRIEZE, ROBBER HOLE,
WINDOW APERTURE, AND SEAM LINE BETWEEN DRUMS

Pocket Landscapes:
Trajan's Monument to Poché

JOSHUA G. STEIN

Standing amidst the cacophony of Rome's Piazza Venezia, Trajan's Column slips easily into the lively frenzy of tourist and city buses, excavation sites (both archaeological and infrastructural), and traffic (pedestrian, auto, and motorino). As anywhere in Rome, these networks of transit, commerce, and artifacts are layered as thickly above the streets as they are buried beneath them. Through this earth rich with aggregate imperial desires, Mussolini carved an axis connecting his Palazzo Venezia office and underground bunker to the Roman Coliseum, revealing and dividing two sets of ancient cellular plazas, the Imperial Fora and the Republican Forum. Along this axis, anchored by the Column, sits Trajan's Market, an imperial complex carved into the side of one of Rome's celebrated hills.

The voids left by both Trajan and Mussolini index a shared scale and intensity of imperial bombast. But while Mussolini's cut through the strata of this city intentionally obliterates certain histories (the medieval) in favor of a singular view of history (imperial conquest), Trajan's Column offers a contrary reading: a simultaneity of perspectives are contained and anchored in this single edifice. Trajan, emperor from 98 CE to 117 CE, remains notable for extending the frontier of the empire to its farthest limits, orienting formerly "barbarian" lands towards Rome. The celebrated Column was erected to commemorate Trajan's conquest of Dacia, now modern-day Romania. This was a calculated act of rational and bloody imperial expansion, completely in keeping with the centralizing politics of Rome. However, the Column also unwittingly acts as a monument to the simultaneity and opacity of place, contrasting sharply with the singular hierarchy implied by the dictum "All roads lead to Rome."

Trajan's Column embodies, epitomizes, and ultimately monumentalizes the contradiction between Rome's desire to locate authority through spatial and political centering and the persistence of the dispersed and unknowable spaces of the city. This sense of the known versus the unknown in urban and architectural space relates to the multiple roles assumed by the architectural condition known as poché. Within

*Nolli's famous
map of Rome
employs graphic
techniques
representing
both material
and conceptual
conditions of poché.*

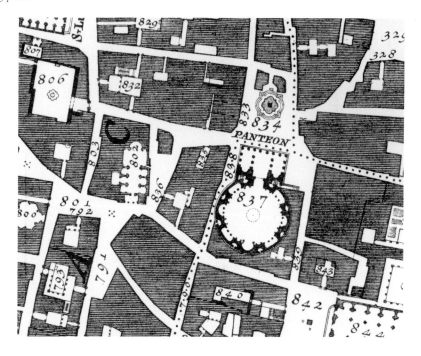

DETAIL FROM GIAMBATTISTA NOLLI'S MAP OF ROME (TOP), 1748,
AND VIEW FROM THE COLUMN'S BALCONY (BOTTOM)

51

the discipline, poché indicates "pockets" of opacity and thickness that separate inside from outside or one room from another. In the most normative conditions, this might mean solid masonry walls. However, nested within this solidity we can also find smaller voids, some of which may be habitable. These void pockets can also be called poché and may include staircases, servant's quarters, or secret corridors which serve the "back of house" needs of the building. These hollows exist simultaneously with the primary voids of the building but are a world apart, mostly unperceived by the public. Because they are hidden, literally and experientially, these poché spaces deny their users an understanding of dimension, geometry, orientation, and ultimately of location.

A third definition of poché refers to a graphic convention within architectural drafting in which closely packed repeating parallel lines are used to infill an outlined form to produce a field of gray tone. The drawing techniques found in Giambattista Nolli's 1748 map of Rome clearly demonstrate these multiple conditions of poché. Nolli's rendering of the Pantheon reveals a differentiation between two different tones of poché. He employs a darker tone to indicate the true solid mass of the Pantheon's twenty-foot-thick masonry walls. The lighter tone he uses elsewhere refers ambiguously to conditions either of material or social opacity—wall thickness versus habitable hollows that are either undocumented or off-limits to the public.

The experience of visiting Trajan's Column shifts one between these different conditions of poché. The tools used to excavate the stairs and apertures within each drum left behind marks that produce a rich effect of textured chiaroscuro upon the day-lighted surfaces of the interior. Wherever sunlight pierces the interior it rakes across the chisel marks, generating an intensely patterned tone much like those produced by Nolli's engraving tools upon his copper printing plates. The Column's richly textured interior micro-landscapes produced by ancient tools echo the mineral landscapes from which the thick stone drums were excavated—the marble quarries near Carrara. More significantly, within this extensively documented city, the interiority of the Column is not only literally hidden from view, but exists as a material lacuna in the consciousness of Romans and tourists alike. Conceptually, the typology of the "column" would normally assign it to the realm of "solid" in the dichotomy of urban mass and voids, denying the possibility of a habitable interior. This reinforces the reading of poché as an unknowable hollow. Trajan's Column nonetheless oscillates in its identity between architecture and sculpture—between fissure and monolith.

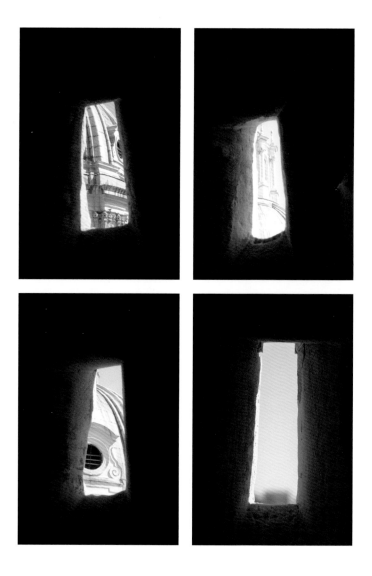

ZOETROPE VIEWS FROM WITHIN THE COLUMN

EXTERIOR VIEW OF WINDOW IN THE COLUMN

The slender Column of Trajan operates as a marker for urban orientation, like the myriad other obelisks, fountains, and statues that populate the center of so many of Rome's piazzas. But it can also be seen as a material condensation of the space defined by its imperial purview—replicating the centripetal tendencies of the Empire, gathering and compressing mass from distant provinces. Trajan's Column shares this quality with the Mausoleum of Hadrian, now the Castel Sant'Angelo. This massive earth and brick cylinder with its internal helical ramp looms over the Tiber. Trajan built his Column, which would become the resting place of his ashes, seventeen years before Hadrian's Mausoleum broke ground but it might be imagined as a dwarf-star version of the latter, condensing all the material of the grand earthen drum while maintaining the central void that houses the body of the Emperor.

With this conceptual density locked within its carved marble drums, Trajan's Column operates as a monument to excavation, and is perhaps the only tower constructed primarily through this process. With the exception of the significant, and much discussed, act of stacking twenty thirty-two-ton drums of marble, the power of this monument is due above all to successive manual subtractions of material. From quarrying the marble, to voiding the internal spiral stair, cutting the forty window apertures, and chiseling the exterior frieze, the creation of Trajan's Column is the fruit of repeated acts of excavation. While the Column is most famous for its intricately carved exterior, the fact that the entirety of its interior was produced by carving, points to much more radical possibilities for architecture. The Washington Monument is the perfect contrast: it aspires to the condition of habitable monument, but its walls, composed of 36,000 stacked blocks of marble and granite, produce a space utterly antithetical to the carved helical grotto of its ancient antecedent.

While the tectonics of assembly seem to connect to location through desire or will, those of carving or excavation are more rooted in material acceptance and exigency, embodying the readiness to work with that which is found. In Gaston Bachelard's subterranean space of the cellar, the act of excavation connects each individual location through the common medium of soil; in the excavated void, the abstractions of geometry, geography, and distance are swallowed by the maw of the earth.[1] What better way to connect the distant conquest of a landscape and people to the immediacy and specificity of the earth in Rome, albeit of a geology originally 250 miles away in Carrara? The connection to the earth through the tectonic operation of excavation produces

EMBRASURE AND TEXTURE WITHIN A COLUMN WINDOW
AT THE JUNCTURE BETWEEN TWO DRUMS

another powerful fiction in which the Column is the form that remains after the earth has been removed from around it, like some sandstone hoodoo in the American Southwest. This reading acknowledges that the Column was intended to reproduce the view from the hill that Trajan demolished to create his Forum and Market (see page 46).

Like so many other archaeological excavations, Trajan's Column creates a thick material buffer against the harsh light and sound of the contemporary world. In this case, rather than descending into the damp must of a historical dig, we spiral upward, simultaneously leaving the earth while becoming more aware of its cool, massive solidity. This projection of excavation out of the earth collapses two different urban perspectives. The view from the top, ostensibly the raison d'etre of the Column, offers a survey of the city, yet the apertures encountered along the journey upward provide no such comprehensive vista. From its interior, Trajan's Column is more of a vertical, perforated tunnel, where each drum operates as a thickened zoetrope, filtering out the gestalt of its context while assembling an animated aggregation of fragments, vignettes, and details. Here the overexposed cityscape of contemporary Rome is glimpsed through a helical mineral sponge. The Baroque domes of Santa Maria di Loreto and Santissimo Nome di Maria and the turn-of-the-century classicism of the Altare della Patria become animated jump-cuts framed by deep marble proscenia.

As the embrasure expands each window aperture from exterior graphic rectangle to capacious wedge of interior space, the pattern of chisel marks highlighted by the oblique light raking across the stone surfaces creates a set of miniature mineral landscapes. As these interior pocket grottoes encounter the exterior bas-relief, itself an illustrated narrative of territory and its acquisition, they maneuver themselves into the gaps between soldiers' bodies, stretching to stand in for a cavalryman's shield, or morphing into the background of the relief's vernacular architecture. At other moments, the specific location of a window aperture in relationship to the assembly of the Column's giant stacked drums creates an intersection of window and seam, one slowly eroding into the next over the millennia. These local "aberrations" produce a set of similarly sized rectangular apertures—each uniquely modified according to its context within the unfolding story of conquest and within the tectonic assembly of the monument—so that an expert scholar of the Column could locate their exact location within this speculative zoetrope based solely on the signature profile of each window aperture.

Rome itself is celebrated by historians both as a center of rational planning and as a subterranean labyrinth of fluid potentialities. The liquid association here is apt, as the massive earthen heterogeneity of this deeply
layered city is due to the walls of Rome operating as a mold into which
the successive layers of material history were poured. The walls trap
and bury artifacts of all sizes within this urban-scaled "cast." The hidden spaces of the city—its ancient aqueducts and sewers, the thickened
double-shelled domes of the baroque churches, and the secretive spaces
of cults—conspire to create an extensive complex of unknowable spaces.
These hidden pockets and labyrinths are the dark reflection of the more
clearly ordered spaces indicated on Nolli's map as white figures. Trajan's
Column constitutes an appropriate, albeit unintentional, monument to
the dual nature of this city—a city that has been studied, interpreted, and
idealized and yet remains persistently thick, opaque, and massy. While the
Column's observation platform surveys and surveils through a rational
understanding of the adjacent cityscape, its material presence connects
us to a less rational underworld extending just below the surface of
our visible existence.

CODA: TRAJAN'S HOLLOW

If my winding history were to continue beyond the original artifact
of Trajan's Column, I would offer an extended tale in which the Column assumes its new role as a monument to poché. *Trajan's Hollow*, a
revisionary reconstruction of one of the nineteen drums of Trajan's
Column, intensifies the dual nature of the original, while reconciling
these contradictory tendencies. Building on the accidental reciprocity
between the interior and exterior of the Column created by its set of
forty apertures, *Trajan's Hollow* develops a formal and spatial language
for communication between the Column's interior experience and the
city that surrounds it. The narrative frieze is transformed into a thick,
yet porous filter through which the poché experience of the Column is
linked to the urban landscape and the epic memory of Trajan's exploits.

  In the original Column, each window is modified by the military
narrative into which it is lodged, creating a deep imprint of narrative
bas-relief projected towards the cylinder's interior. *Trajan's Hollow* applies
this logic to the entire exterior surface, amplifying the texture of the
sculptural frieze to such extremes that it calls into question the purity of

*This revision
of one drum
of the
original Column
generates a new
pattern of apertures
linked to the
narrative frieze.*

the original geometry as well as the harsh distinction between interior and exterior. The delicate bas-relief, originally painstakingly chiseled to a depth of 3 CM into the marble, is instead excavated to over ten times this depth, to the point of exceeding the 30 CM thickness of the drum itself. These cuts break through to the very interior that the original scroll once masked from the sight and consciousness of the city. In carving through the thickness of the Column wall with the same rhythm of its original narrative, the bas-relief account becomes equally perceivable from the perspective of its interior, creating a new set of fifty or more apertures per drum where once there had been only three. Through this greater perforation, a new zoetrope of the emperor's exploits in Dacia is created at a much finer resolution than the original, and the carvings of the bas-relief are transformed into truly spatial and experiential conditions. Here, glimpses of the Roman cityscape around Piazza Venezia may be assembled via the story-arc of Trajan's conquest of Dacia, a narrative translated into sequences of measured negative space.

Trajan's Hollow proposes an alternative reading of Trajan's monument, departing from the physical artifact left to us, which exists quite independently from the desires of its architect, Apollodorus of Damascus. While the narrative frieze and the vertical form proclaim the triumph of imperial conquest, its interior offers the contrary experience of the pockets of culture and space within the city and empire that have escaped totalizing surveillance. The kingdom of Dacia as a distinctive cultural region was obliterated, or at least buried, by the overlay of Roman military and political conquest. But like the many cults and cultures supposedly assimilated into the empire, the spatial complexity inhabited by these subjugated populations remains in the subterranean spaces of Rome. These cultures quietly inhabit the city as do the poché spaces of a building. The Column not only offers a survey of the city from above, but also reproduces the perspectives from within the pockets of thick or underground spaces below. A creative misreading of Trajan's Column introduces a porosity to the Roman notion of imperial space, now infiltrated by a rhizomatic network of experience. By complicating the dichotomy of inside versus outside (so necessary for governance by imperial dictate), and above ground versus below, the Column offers a simultaneity of idealized abstractions and specific material events.

An earlier version of this essay was published in *Floor: A Journal of Aesthetic Experiments* no. 2 (2013).

1. Gaston Bachelard, *The Poetics of Space,* trans. Maria Jolas (Boston: Beacon Press, 1964), 18–20.

Hidden Trajan

HIDDEN TRAJAN

While the Column has been meticulously studied by historians, archaeologists, and architects, there remain significant lacunae in its documentation. It is as if entire narratives remain hidden from both the public and scholarly inquiry. Research and analysis of the actuality of the Column (at any particular point in history) have been eclipsed by theories and speculations concerning Apollodorus of Damascus's intentions and his original designs for the Column. While the most overarching oversight is the existence of the Column's interior, a number of key observations about the nature of the Column follow from this gap in research.

Before my visit to the Column, my research was primarily concerned with presenting the existence of the interior. After the visit, I was most struck by the need to examine three specific aspects revealed from this interior perspective:

1. Edge Conditions
The junctures between each marble drum visualize the conflict between design aspirations and material affordances.

2. Outside In
The slow erosion of the Column frieze creates an unexpected material link to its interior.

3. Mineral Sponge
The specific profiles of each aperture provide the key to connecting the interior experience with the external narrative frieze.

TEXTURE ON SURFACE OF COLUMN'S CENTRAL CORE

*Piranesi represents
the idealized
drum along with
the system of
connections to its
neighbors above
and below.*

EDGE CONDITIONS

While Piranesi's sectional drawings depict clean seam lines where each drum connects with its neighbors above and below, this is not the condition one finds upon visiting the interior of the Column. Instead, the juncture between each drum results in a consistent rupture where the acute angle of the soffit breaks at each connection point. As the underside of the helical soffit meets the stepped surface of the spiral stair, the marble is asked to maintain a razor-thin edge which exceeds the capacity of the material.

It is very unlikely that Piranesi witnessed the smooth drum connections that he drew in the 18th-century (what are the chances that all of this material would have remained intact for seventeen hundred years only to spall off in the last 250?). In fact, it is entirely possible that these ruptures were a condition created during the Column's original construction. We could imagine the bottom segment of each drum's stairs breaking off at each moment the thirty-ton mass "settles" into heavy contact with the drum below. If this were the case, did Apollodorus of Damascus witness the marring of all but two of his massive drums? Did he regret his design? Or perhaps his choice of material? While it seems impossible to consistently stack drums made of Luna marble without chipping off any acute angles, there are certainly other materials that could have handled this geometry and assembly technique.

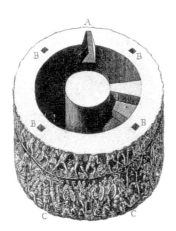

G.B. PIRANESI, "ILLUSTRATION OF THE DRUMS AND HOW THEY WERE CONNECTED" (DETAIL),
PLATE VIII, 1774–75

The smooth and barely perceptible "ideal" meeting of two drums is the exception that proves the rule: monolithic spiral stairs cannot be cut and stacked out of Luna marble without compromising the material integrity of the stone.

While his possible intentions and reactions are interesting, what is more important for our history is the realization that we have collectively dedicated so much time imagining the idealized designs of the ancients that we have ignored the actualities of their artifacts as presented in their current state, which may hold other lessons concerning materiality and technique.

THE EXCEPTION: SMOOTH DRUM CONNECTION

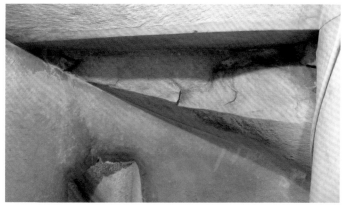

In all but two instances (16 out of 18 drum connections), each drum chips off at approximately the same location. The marble becomes too weak to hold the infinitely thin profile of the stairs as they meet the smooth helical soffit on their underside.

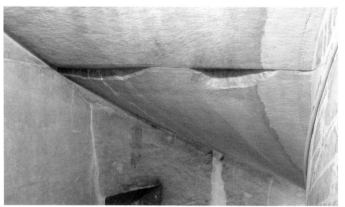

THE RULE: TYPICAL DRUM CONNECTIONS

*The Italian
peninsula slowly
turns itself inside out
as the Apennines
erode through the
Tiber into the
Tyrrhenian Sea.*

OUTSIDE IN

It is well-known (to some at least) that the 19th-century plaster casts of Trajan's Column, protected from the elements in various institutions, have preserved the fine detail in the sculptural relief far better than the original Column itself. Less considered is an examination of where the eroded marble from the original has gone over the millennia. We assume most of this material has simply found its way from the exterior of the Column to the city's gutters, the Tiber, and ultimately the sea, contributing to the slow process of the Italian peninsula turning itself inside out as the mountains become the sediment that continually adds to the coastline (Rome's ancient port Ostia is now separated from the coast by 9 kilometers of land). However, a certain percentage of the Column's eroding bas-relief has instead found its way through the Column's windows, passing from the exterior surfaces to those on the interior. The Column has become its own storage vessel.

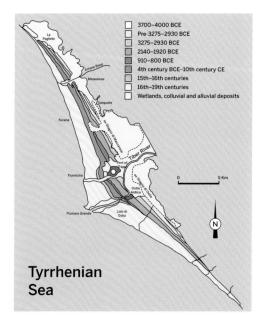

☐	3700–4000 BCE
☐	Pre-3275–2930 BCE
☐	3275–2930 BCE
☐	2140–1920 BCE
■	910–800 BCE
■	4th century BCE–10th century CE
☐	15th–16th centuries
☐	16th–19th centuries
☐	Wetlands, colluvial and alluvial deposits

Tyrrhenian
Sea

TRANSFORMATION OF THE TYRRHENIAN COASTLINE PRODUCED BY
TIBER RIVER DELTA DEPOSITS OVER SIX THOUSAND YEARS

*Over the centuries,
the Column
has become the
container for its
own dissolution.*

Could Apollodorus of Damascus have anticipated this transference of material from the outside to the interior? The Romans knew about the deposits created by calcium-rich water and designed their aqueducts with maintenance walkways that allowed the periodical removal of this build-up. While historians have spent centuries debating the idealized systems of geometry developed by the ancients, could we not learn an equal amount about issues beyond proportion and composition? How might one develop drainage systems that anticipate the slow erosion of our buildings? How could a building operate as a vessel for its own dissolution?

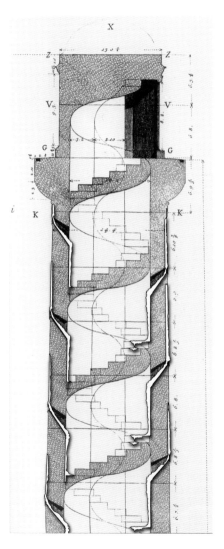

DIAGRAM OF COLUMN EXTERIOR ERODING INTO ITS INTERIOR

SUPERIMPOSED OVER G.B. PIRANESI'S,

"VERTICAL SECTION OF TRAJAN'S COLUMN" (DETAIL), PLATE IV, 1774–75

CALCIUM DEPOSITS FROM ERODED EXTERIOR CASCADE
DOWN THE SPIRAL STAIR OF TRAJAN'S COLUMN

As the exterior frieze erodes, it drains through the windows, cascading down the stairs— the Column turns itself outside in.

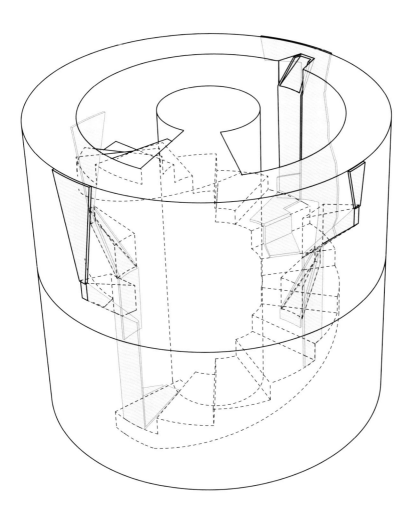

MAPPING OF TRANSFER OF CALCITE MATERIAL
FROM EXTERIOR OF COLUMN TO INTERIOR

Evidence of the drainage and calcification patterns exist alongside the tests by preservationists to remove this build-up.

EXAMPLES OF ACCRETION WITHIN THE COLUMN

MINERAL SPONGE

Although the forty windows of the Column shaft are barely visible from the exterior, from the interior they dominate the experience of ascending. Embrasure, the splay of the window from exterior to interior, creates a surprisingly generous wedge of space if not view. Perhaps more importantly, these windows create a link between the exterior frieze and the interior experience. Each aperture is unique, negotiating the specifics of the narrative on the exterior with the space of habitation on the interior as well as the juncture points between the massive drums.

The Column's window apertures operate as a conduit for the movement of information from the exterior to the interior. In addition to the penetration of the outside city or the flow of material from outside to inside, the exterior narrative content also infiltrates inward. The Column's bas-relief frieze, its most celebrated aspect, projects into its interior, although only as miniature episodes.

Because each aperture is uniquely modified by its context in the narrative frieze and drum construction, one could imagine that spiraling up the interior stairs, the perfect scholar would be able to locate their position within the frieze each time they happen upon a window.

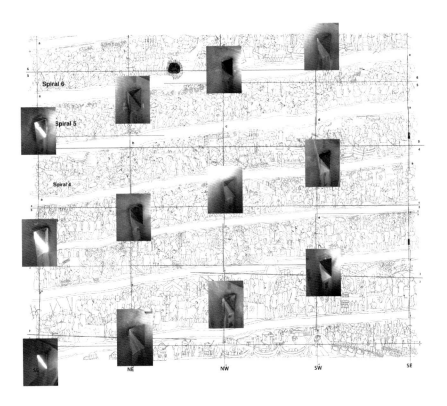

Spiral 6

Spiral 5

Spiral 4

SE   NE   NW   SW   SE

ALIGNMENT BETWEEN WINDOW APERTURES AND NARRATIVE FRIEZE

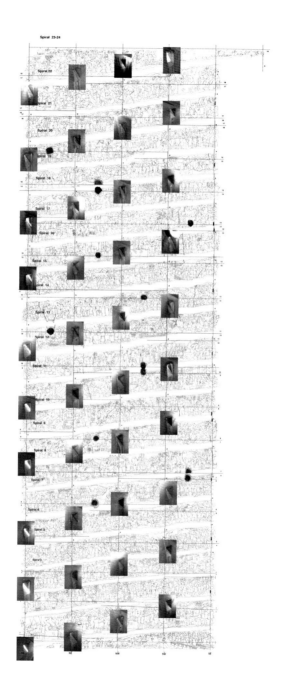

WINDOWS MATCHED TO LOCATION ON EXTERIOR FRIEZE

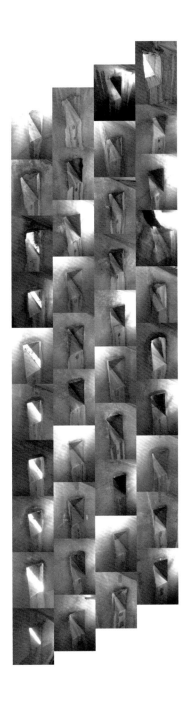

MATRIX OF WINDOWS AS VIEWED FROM COLUMN INTERIOR

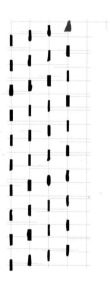

ZOETROPE VIEWS FROM WITHIN THE COLUMN

Images by Michael J. Waters

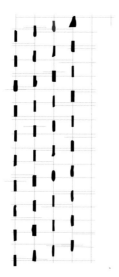

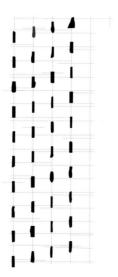

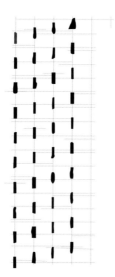

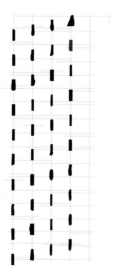

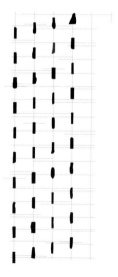

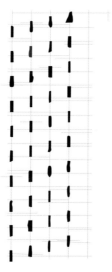

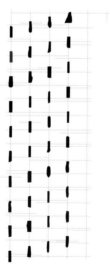

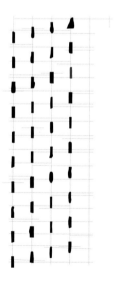

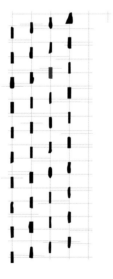

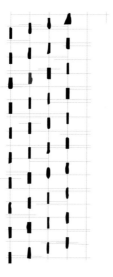

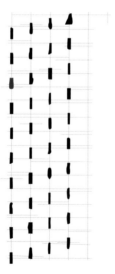

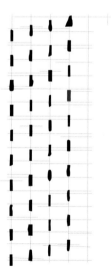

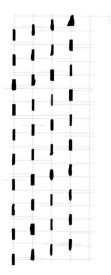

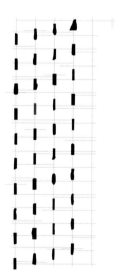

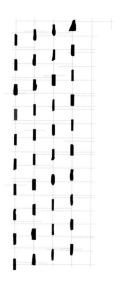

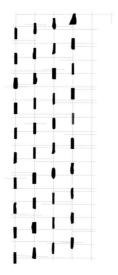

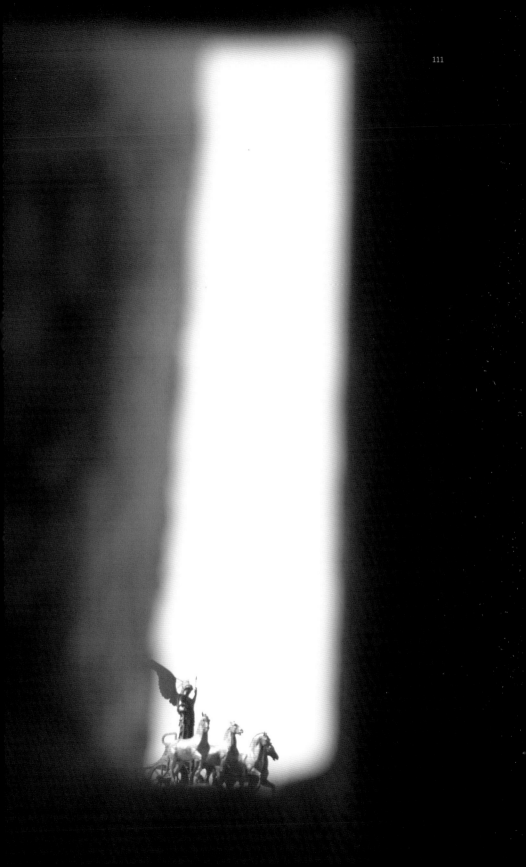

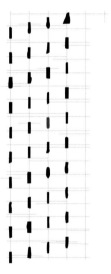

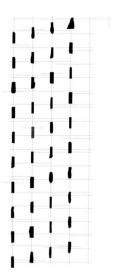

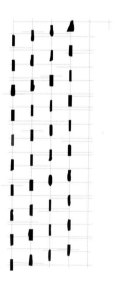

Trajan's Progeny

Imitations

History as a model to move towards

Monuments to the past

Aspiring to the greatness of Trajan's Column

Imprints

History as a source to move from
Aberrant materiality of the past
Claiming direct connection to Trajan's Column

IMITATIONS VERSUS IMPRINTS

A fundamental distinction can be drawn between two very different lineages in Trajan's progeny. On the one hand, a series of imitations aspire to the greatness of the original, mimicking its profile. On the other, a tradition of imprints use the original as a site for material sampling, launching into more radically diverse interpretations of the historical original.

Trajan's
Imitations
spiral
around
an
established
set
of
tropes
while
codifying
the
Trajan's
Column
type.

Column of
Marcus Aurelius
Rome, Italy
176–193 CE
Spiral Stair
Monumental Statue
Narrative Frieze

Karlskirche
Columns
Vienna, Austria
1716–1737
Spiral Stair
Narrative Frieze

Vendôme Column
Paris, France
1806–1810
Spiral Stair
Monumental Statue
Narrative Frieze

*Washington
Monument
Baltimore,
Maryland, USA
1815–1829
Spiral Stair
Monumental Statue*

*Congress Column
Brussels, Belgium
1850–1859
Spiral Stair
Monumental Statue*

*Astoria Column
Astoria,
Oregon, USA
1926
Spiral Stair
Narrative Frieze*

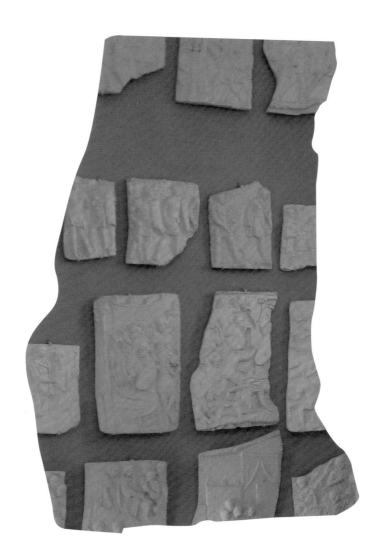

Trajan's
Imprints
claim
a
material
pedigree
linked
to
Trajan's
Column,
but
are
free
to
drift
from
this
model.

Académie de France
Villa Medici
Rome
1665–70
Trophy Fragments

Victoria and Albert
Museum
London
1861–62
Halved Hollow
Shell

Museo della
Civiltà Romana
Rome
1861–62
Unravelled
Narrative

American Academy
in Rome
Rome
2010–2011
Horizontal
Cross-Section

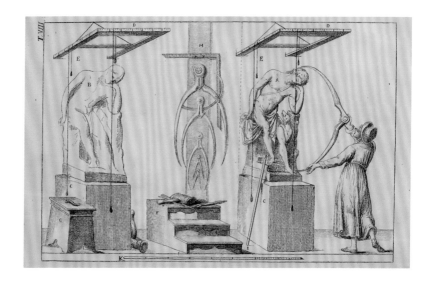

FRANCESCO CARRADORI, "RULES FOR CARVING ANY SCULPTURE FROM MEASUREMENTS" FROM
ELEMENTARY INSTRUCTIONS FOR STUDENTS OF SCULPTURE, 1802

The Wayward Cast:
Gipsotecas, Digital Imprints, and the
Productive Loss of Fidelity

JOSHUA G. STEIN

The speed with which imitations can now be produced (sometimes appearing even before their "originals"[1]) may soon render direct mimicry obsolete or simply uninteresting. In his account of the loss of aura, Walter Benjamin notes the discrepancy between the time-intensive methods of manual reproduction versus those by machine: the quickness of the eye replaces the labor of the hand.[2] But while speed may be a primary factor in this transformation, materiality is equally at play. An investigation into the traditions and techniques of physical transference of qualities from model to copy may offer a more "productive" form of reproduction than the visual imitation—one that offers new perspective on the original while escaping the tyranny of its likeness. Rapid advances in digital scanning and surveying technology in archaeology, historical preservation, and geomatics offer means of reproducing formal information that are no longer tied to the visual, our primary mode of reproduction. Architecture will need to quickly redefine its default methods of historical analysis and its role in contemporary design.

The production of copies was not always understood to be an act of forgery. Ancient Rome's ability to borrow from other cultures is well-known, including the assimilation and translation of Greek and Etruscan tropes and forms by Roman artists and architects. But there existed a parallel tradition—the less intellectualized process of analog reproduction that yielded duplicates of the past initially intended more as documentation than inspired variation patterned on historical models. This included a respected craft tradition in the creation of plaster casts, objects that for several centuries were highly regarded and collected. Although currently these reproductions tend to be associated more with artisanal or archaeological traditions than fine art, they offer another paradigm by which we might understand a contemporary possibility for the copy.

CASTS OF TRAJAN'S COLUMN AT THE VILLA MEDICI (ACADÉMIE DE FRANCE À ROME)

TRAJAN'S PROGENY:
IMITATIONS VERSUS IMPRINTS

Trajan's Column in Rome, perhaps because of its synthesis of monumental architecture and exquisite narrative bas-relief, has spawned a
continuing trajectory of impersonations and facsimiles. It quickly becomes important to distinguish between the mimetic endeavors that
view the Column as inspiration versus the casts made of its exterior
that were intended to produce precise replicas of Column's surface. Of
the former, there are numerous examples hailing from as early as the
Column of Marcus Aurelius completed in 193 CE, just eighty years after
Trajan's. Located a mere 700 meters away, this imposter is often confused
with the monument on which it was modeled.

Many other prominent examples of imitations followed, stretching
from antiquity into the 20th-century, each a slight variation on Trajan's
theme. The creators of these likenesses favored a relatively faithful reproduction of the imposing profile of the Column plus an interior spiral
stair, an exterior narrative frieze, a massive rectangular pedestal and a
statue topping off the column. These tropes characterize the Column
of Trajan type, a type which serves as a wax cylinder ready to record
any new display of monumental power. In each new iteration, the statue
of Emperor Trajan is removed (usually to be replaced by another all-
powerful figure), and the exterior narrative rewritten.

But there is a divergent history of copies of Trajan's Column that
offers a more aberrant and productive tendency: that of the casts. This
lineage is perhaps even more fascinating in its desire to capture the
authenticity of the original through the material transference of its
qualities—formal, textural, procedural. In his catalog to the exhibition
L'Empreinte, art historian and philosopher Georges Didi-Huberman dif-
ferentiates between duplications made through the regime of the image
(imitations) and those through the regime of matter (imprints). In his
essay "La Ressemblance par Contact," he positions the imprint as an
archetype of thought and sensibility as much as technique.[3] The imprint
allows for a tactile (as distinct from visual) transmission of qualities from
the original to the copy, dragging with it all the "imperfections" of the
physical world that the eye might edit away. While Didi-Huberman
lists a range of techniques that might be understood as imprints, the
cast is one of the most obvious examples. To produce a cast demands
an intimate negotiation with material tendencies, focusing attention

towards the process of making and away from questions of intention or semiotics. In effect, it denies the distance necessary for what we consider intellectual or artistic license. Although the Column has been the subject of reproductions that fall along trajectories of both the imitation and the imprint, it is the weighty history of casts that best distinguishes the Column from other monuments while holding the most promise for contemporary practice.

FRENCH REPETITION OF REPRODUCTION

It is the French who initiate and channel the flow of plaster into and out of Rome (see page 227). For obvious reasons, the French monarchs and dictators would maintain a continuous desire to symbolically and materially transpose the glory of the Roman Empire to Paris, resulting in the waves of casts of Trajan's Column commissioned over three major campaigns—François I (1540), Louis XIV (1665–70), and Napoleon III (1861–62). Napoleon Bonaparte, who had hoped to physically move the artifact of Trajan's Column itself to Paris, instead contented himself with an imitation erected in the Place Vendôme. But his desire to possess the relics of antiquity, or at least their likenesses, was responsible for the great flow of artifacts toward the Empire's capital. It was the 1861 campaign by Napoleon III that yielded the most thorough (and productive) set of molds of the Column. This operation was a truly modern endeavor that created not only the most extensive replica of the Column, but also a machine for its material propagation. The molds were produced through the newly invented process of electrotyping (or galvanoplasty), which generated a durable metal mold capable of producing multiple casts. These molds, housed in the Musée du Louvre in Paris, have been used to manufacture a small (and slowly growing) diaspora of casts. Complete plaster sets exist in the Victoria and Albert Museum in London and the Museum of Roman Civilization in Rome with another "recently" produced for the National History Museum in Bucharest in 1968.[4]

The Cast Court of the Victoria and Albert Museum in London, one the most famous cast galleries, is anchored by a replica of Trajan's Column. This cast is assembled erect but partitioned into two segments, castrated by the confining height of the Cast Court's glass roof. There could be no better indication of the valued qualities to be presented to the British public: symbolic power of size. In this version, the most intact

*In the V&A casts,
the obsession with
surface narrative
trumps any attempt
to exhibit a sense
of the mass, space,
or material
of the original.*

INSTALLATION OF THE CASTS OF TRAJAN'S COLUMN (TOP)

AND ITS INTERIOR (BOTTOM)

AT THE VICTORIA AND ALBERT MUSEUM, LONDON

*This V&A's
severed cast
of the Column
sits among
the menagerie
of other historical
samples on display
in the
Cast Court.*

CAST OF TRAJAN'S COLUMN DISPLAYED IN THE CAST COURTS

AT THE VICTORIA AND ALBERT MUSEUM, LONDON

physical representation of the Column, there is no attempt to offer any understanding of the interior space (which would of course be difficult given that casts were only ever produced of the exterior). The fragmentation of the Column echoes the curatorial agenda usually associated with the museum cast gallery—a pastiche of artifacts providing the public the opportunity to grasp a "best of" sampling from across historical styles and epochs.

THE MONUMENT UNRAVELED:
THE CAST AS RECONTEXTUALIZATION, AS REREADING

In contrast to the V&A's reconstruction, the plaster copies of Trajan's Column in Rome's Museum of Roman Civilization exemplify the potential of the cast to redefine or reread its original referent. Located in EUR, the fascist exurban city created for the 1942 Esposizione Universale di Roma, the Museum of Roman Civilization was Mussolini's attempt to gather in one location representations or reproductions (depending on the size of the original) of all the great statuary and monuments of the Roman Republic and Empire. The museum attempts to create a chronological journey following the triumphs of Rome, ultimately intended to serve as a justification of Mussolini's own imperial aspirations. In direct opposition to the V&A's Cast Courts, the Museum of Roman Civilization presents information didactically and diachronically as it traces the development of one particular culture over history.

Because the museum houses statuary, architectural elements, and dioramas that are intended to display an anthropological narrative, all artifacts contained within are meant to be viewed as the originals would have appeared in antiquity. Any marks from the casting process are sanded away, and whether plaster casts or fiberglass resin casts, all the statuary is painted with a patina to simulate the original marble or stone (a clear conflation of Didi-Huberman's imitation/imprint dialectic). The display of Trajan's Column sets up a conflict between the two desires of this institution—on the one hand to recreate the appearance of the original artifact and on the other to present history through chronological narrative. Here, Trajan's Column is arranged as a series of separate panels, as if the bas-relief frieze were unrolled to produce a continuous 200-meter-long plaster scroll, folded into four lengths to fit within a corridor connecting two wings of the museum. While the imposing profile of the Column cannot be appreciated (as the V&A attempts), this arrangement offers one of the only opportunities to read the frieze as a continuous and complete narrative (something never possible in Trajan's original). This presentation of Trajan's Column capitalizes on the potential of the cast to extend and elucidate one reading of the original by abandoning the desire to provide a mimetic visual experience.

Since the bulk of the other casts at EUR masquerade as ancient statuary, they depend on their visual likeness to their originals and in this way, despite being created through the technique of casting, might be more comfortably classified as imitations rather than imprints. By contrast, the casts of Trajan's Column rely more on their provenance—their extension of the original through the material transference of its qualities. This exoneration of the necessity for visual likeness in favor of the pedigree of process is typically more prevalent in the cast galleries or gipsotecas contained within fine arts academies where the process and material of plaster casting offers a desirable acknowledgement of the copy as something other, perhaps even something more, than the original. It enables and encourages the indexing of the various techniques of reproduction onto the inherited form. The matrix of seam lines becomes the embodiment of the casting process, deepening the connection to the original while simultaneously eroding its visual similarities. In this species of cast gallery, the desire to study and explain the original leads to an openness to its transformation through reproduction.

THE WAYWARD CAST:
THE MISPLACED FIDELITY OF TRAJAN'S HOLLOW

Trajan's Hollow 1.5, a reconstruction of a cross-sectional "slice" of Trajan's Column, continues the trajectory of plaster casts of the monument, accelerating the mutation facilitated through the mode of the imprint. Like the set of casts at EUR's Museum of Roman Civilization, *Trajan's Hollow* radically departs from its original in terms of its massing, replacing the vertical profile of the imperial Column with the landscape of a fissured horizontal ring—a five centimeter-thick cross-section of the original Column produced as a floor cast at 1.5 times the scale of the original. While the EUR casts deconstruct the Column's profile and mass in order to reconstruct its historical narrative, *Trajan's Hollow* instead sheds the celebrated external surface in favor of solidifying the original's

The dismemberment and sequential arrangement of the cast segments allows one to read the complete narrative in a manner that the original column could never offer.

DETAIL (TOP) AND DISPLAY (BOTTOM) OF CASTS OF TRAJAN'S COLUMN AT THE
MUSEUM OF ROMAN CIVILIZATION IN EUR, ROME

thickness and its architectonic and spatial qualities. This new cast favors a rendering of the mass of the Column walls and helical void over the sculpted surface of the bas-relief exterior (a perfect inverse of the shell cast housed at the V&A).

Given the historical catalog of other casts of the Column, each of which advances a different rereading of the original, this reproduction explores the opportunity to focus myopically on a singular and over-looked aspect of the original: the relationship between its interior and exterior. In *Trajan's Hollow*, there is a rigorous focus on technique (in particular fabrication, craft, and scalar shift) followed by the freedom to examine only one aspect of the Column, cut loose from the responsibility of capturing its entirety. The symbolism of the Column (so essential in all of its imitations) becomes intentionally subverted through this act of reproduction.

Trajan's Hollow's extreme divergence from its referent while remaining connected through its material genesis advances the notion of the "wayward cast." While the typical copy strains to achieve ultimate fidelity to the original while simultaneously inhabiting the impossibility of realizing this endeavor, the wayward cast attempts to reconcile this paradox by allowing process to derail the reading of the original even more completely. Counter to the larger aesthetic of the Museum of Roman Civilization, *Trajan's Hollow* permits the residue of casting to remain, constructing an entirely different reading of the original. In *Trajan's Hollow*, the plaster that squeezes between the mold seams remains, calcifying into a textured rhythm of fins in syncopation with that of the intensified bas-relief, slowly redefining the original.

THE DIGITAL IMPRINT:
MAPPING SURFACES AND MINING FISSURES

While *Trajan's Hollow* is rooted in the casting lineage of Trajan's Column, it is also one of a series of contemporary architectural exercises investigating the potential of reproduction through the virtual "contact" offered by new digital scanning and fabrication techniques. Photogrammetry, lidar, 3D laser scanning, and other forms of remote sensing may at first seem counter to the intensely physical, analogue techniques that Didi-Huerman associates with the imprint. However, their predilection towards surface-based mapping privileges a type of virtual haptic

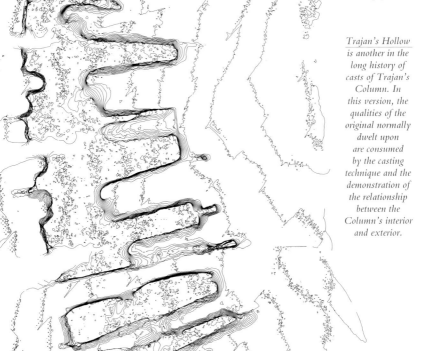

Trajan's Hollow is another in the long history of casts of Trajan's Column. In this version, the qualities of the original normally dwelt upon are consumed by the casting technique and the demonstration of the relationship between the Column's interior and exterior.

DIGITAL PHOTOGRAMMETRY SCAN (TOP) AND DETAIL (BOTTOM)
OF *TRAJAN'S HOLLOW 1.5*
AT THE AMERICAN ACADEMY IN ROME (2011)

*Botta's temporary
one-to-one facsimile
of Borromini's
San Carlo alle
Quattro Fontane
recasts a portion
of the original
in a radically
different tectonic.*

MARIO BOTTA'S *SAN CARLINO LUGANO,*
CONSTRUCTED 1999, DEMOLISHED 2003

over the visual to create what could be called a "digital imprint." This meticulous surface scan is the antithesis of a gestural sketch—more akin to a blind contour drawing. There is no quick summary or snapshot, only textural and topological qualities or data, obsessively traced yet devoid of intention or interpretation. These technologies read a form's surface much as plaster would carved marble (or as the electroplating techniques of Napoleon III's molds of Trajan's Column), blindly tracing surface through multiple particles or waves. While these techniques document an incredible degree of textural definition, "view shadows" reveal blind spots where form eclipses information as undercuts do within the casting process. These fissures in surface data require invention and transformation within the process of reproducing an original.

In 1999, to celebrate Borromini's 400th birthday, Mario Botta constructed a "model" of Borromini's San Carlo delle Quattro Fontane. While the original was in Rome, this reconstruction was erected in Lugano, Switzerland near Borromini's birthplace. This reproduction attempts to "faithfully" document the original form at one-to-one scale. However, rather than rebuilding in stone, the new *San Carlino Lugano* is instead fabricated out of wood: 35,000 planks, each 4.5CM thick, are skewered by steel cables leaving a 1CM gap between each wooden course. This simple swap of material tectonics creates a bifurcation that spawns a completely divergent reproduction. But what enables this divergence is the connection to the original through the photogrammetric survey of Borromini's complex dome and interior. This digital mapping of the complex baroque geometry offers the ability to produce a precise dimensional replica of the original that, while sharing little visual resemblance in terms of architectural massing, façade, or plan, maintains a pedigree through the digital imprint of the original.

Trajan's Hollow, in addition to holding material associations through plaster casting, claims its provenance through an equally intricate technique of digital indexing. Although the digital model was "constructed" as opposed to "mapped," this geometry was merely scaffolding for a parametric translation of the bas-relief figuration into instructions for CNC fabrication, resulting in a series of plaster crenellations. Behind the visually legible correlation of this process— each figure on the original narrative frieze produces one carved void in the reproduction—exists a complex digital rigging inaccessible to the public, shrouded in the technique of the digital imprint. This ensures an authenticity, albeit opaque, towards the original despite the visual discrepancies.

CURATED SYNECDOCHE:
RECONTEXTUALIZING THE IMPRINT

With this wayward tendency of the imprint comes an increased importance on its contextualization, curation, and presentation. *The Villa Rotunda Redux*, produced by British architecture firm FAT (Fashion Architecture Taste) for their Museum of Copying at the 13th Venice Architecture Biennale in 2012, is a scaled-down foam cast of one fourth of the Palladian original (or rather a simplified digital version found in an online archive) presented in tandem with its mold so that the two together imply the mass/void of the full rotunda. In this example, the digital imprint, however simplified, authenticates this reproduction while relieving it from the responsibility of absolute fidelity to the original. Instead, the requirements of the material and fabrication technique, that of volumetrically casting foam against a CNC-fabricated mold, are allowed and encouraged to transform the original model. The charged void between mold and cast, where the Palladian orchestration of interior volumes should play out, is instead a space of vestigial effects from the casting process. Its partial completeness offers the space for critique of contemporary imitation, a commentary not only on the Villa Rotunda, but on the entire lineage of its reproduction: Palladianism and Neo-Palladianism—Home Depot and all.

In Botta's *San Carlino Lugano*, only one half of the original chapel is produced, leaving a full-scale physical rendering of the architectural section drawing—the graphic poché materialized as black paneling on the copy's new face. Although not actually the case, the half-form initially appears to be a giant mold that might be used to produce an architecturally scaled cast. *Trajan's Hollow* also relies heavily on synecdoche: through the association of plaster with material transference, the piece claims to actually be Trajan's Column, albeit only a thin slice. This sampling of the Column, modulated, enlarged, and displaced, still connects to its material origins in the marble of the original.

Each of these contemporary "casts" is intentionally excerpted (as two quarters, one half, one chunk/slice). These partial reconstructions produce a simultaneous reading of both architectural exterior and interior—impossible to apprehend in the original construction. It is as if they reinvent Colin Rowe's notion of literal transparency through their lack of completion. These fragmentary samplings both imbue the original with new information and exonerate the copy from

This cast of one quarter of Palladio's original plus its corresponding mold is arranged to create a doppelgänger in foam and steel.

FASHION ARCHITECTURE TASTE (FAT)'S *VILLA ROTUNDA REDUX*
AT THE 2012 VENICE BIENNALE

the responsibility of complete visual imitation. As in the reliquary and the gipsoteca or cast gallery, the method of display and interaction for a contemporary audience becomes as important as the artifacts themselves. While both the imitation and the imprint were historically a part of the minor arts or trades, Marcel Duchamp's incorporation of the latter into his critical practice, highlights its potential to inform avant-garde production. In his use of molds and casts, the imprint carries weight both through its material connection to the original and through the act of its recontextualization. While the fabrication techniques of the wayward cast may seem errant, its strategies of excision and siting are precise and premeditated.

ARCHITECTURAL IMPRINTS: BUILDING-SCALED CASTS

As emerging scanning techniques continue to become increasingly integrated into architectural practice, the accompanying new data offers the prospect of an architecturally-scaled cast through the digital imprint. While the imitations of Trajan's Column are condemned to reshuffle a limited set of tropes that continually return to the image of imperial power, this contemporary version of the imprint escapes this feedback loop of symbolism. Instead of a "monument to," the casts propose a "piece of" the past—a much more dynamic and productive use of the historical that is continually open to reinterpretation. While the strategy of collaged citation is so (in)famously associated with the post-modern, the scan and imprint achieve variation and relevance through a myopic attention to the exigencies of material process coupled with a transformative recontextualization of the copy.

The contemporary examples above prove most productive when they attempt to literally reproduce form through vastly different techniques of facture from their originals, forcing the final material rendering to stray from any larger visual fidelity to their referents. This is then furthered by a strategy of synecdoche or selective sampling of the original. In addition, none of the three examples ever purports to be architecture, presented instead as artifacts on display in galleries and pavilions. But although they never attempt to reproduce the function of their source architecture, their potential for programmed architectural space feels imminent.

What is most important is how these examples escape the most common critiques leveled against post-modern references and pastiche. New scanning and indexing techniques, as plaster casting did before, ignore the intentions of the original artists and their duplicators, focusing instead on the actuality of their artifacts—blemishes, distortions, and all. They position historical artifacts not as idealized geometric fantasies or ruins for contemplation, but as data to be "unzipped" into a contemporary context; an unearthing of the past not for facile citation or imitation through visual likeness, but rather a transformation of information through contemporary process, technique, and ethos. This notion of history as source material also activates an immediate past that can be recursively repositioned, recontextualized, and re-rendered. This could foster a new species of deviant archaeology where wayward casts are allowed to cultivate a continual revealing of history through the act of reproduction, sidestepping the contemporary western preoccupation with the discrepancies between the authentic original and its "substandard" copies.

1. One of the most infamous contemporary cases of illegitimate copying involved Zaha Hadid's Wangjing SOHO complex in Beijing which was plagiarized by a pirate construction firm in Chongqing. This modified design was possibly obtained as a digital file of an earlier version of Hadid's design. The scandal broke in 2012 as the counterfeit version threatened to complete construction before the original.

2. Walter Benjamin, "The Work of Art in the Age of its Technological Reproducibility," in *Walter Benjamin: Selected Writings*, Volume 3, 1935–1938, ED. Howard Eiland and Michael W. Jennings, trans. Edmund Jephcott et al. (Cambridge, MA: Belknap of Harvard University Press, 2002), 102.

3. Georges Didi-Huberman, "La Ressemblance par Contact: Archéologie, Anachronisme et Modernité de l'Empreinte," in *L'Empreinte* (Paris: Les Éditions du Centre Georges Pompidou, 1997), 69.

4. Valerie Huet, "Stories One Might Tell of Roman Art: Reading Trajan's Column and the Tiberius Cup," in Art and Text in Roman Culture, ED. Jas Elsner (Cambridge: University of Cambridge, 1996), 13.

Part II—Reconstructions

A Set of Directions for the Reader: 185 Steps

MICHAEL SWAINE WITH ILLUMINATIONS BY ARCANGELO WESSELLS

ONE

If it had been possible to build the Tower of Babel without ascending it, the work would have been permitted. FRANZ KAFKA

TWO

Count the number of windows as you walk up the steps.

THREE

As I type windows it feels wrong; the right term should be "hole" or "empty space." A hole feels like we were there with a pick or chisel, digging, chipping away, making a tunnel toward the light or the air. Count the number of holes in this tower.

FOUR

Step two, step three, step four ... then a window, the light hits our face. And again, step, step, step, step, window—and then we start to count the steps between the windows. Then we get absorbed in the pattern of steps and moments of light. We look out the hole.

FIVE

The view "takes our breath"... our breath is taken away, that is the expression. To take one's breath away. As the light warms our face, we take a breath. You and I are on the inside of this helix, spiraling up, breathing together.

SIX

My friend Albert, when I was young, read in a training book for track and field that when training you can hold your breath while running, to increase the ability for our bodies to work without oxygen. We would go to the track and do sets of sprints while holding our breath. We would run halfway around the track, so we held our breath for 1/8th of a mile. The Romans used a measurement distance called a stadium. Roman roads had rocks marking these stadia, eight rocks in a Roman mile.

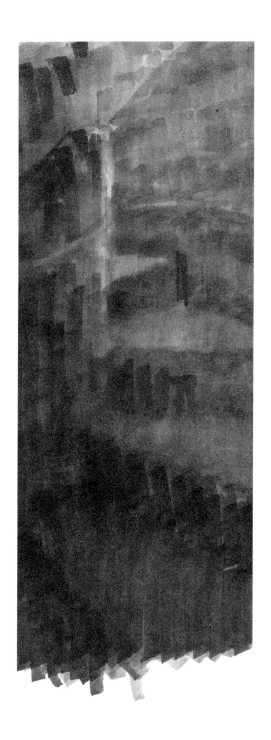

SEVEN

I have my father's calculator. It is called an Otis King Calculator. It is a type of slide ruler, where the line of numbers has been wrapped around a cylinder in a spiral. This creates a more accurate tool. It is like having a slide rule that is 60 inches long. With this calculator I have done some calculations.

EIGHT

A few notes on measurements used during the time in which Trajan's Column was made. Most measurements were and still are related to the foot (pes). A thumb (uncia) is 1/12th of a foot (pes). A finger (digitus) is 1/16th of a foot. A stade is 625 feet (pedes) or a stade could be translated into 10,000 fingers (digitus). The Egyptians divided the Earth's circumference by 60, three times (60 × 60 × 60): that equals one stade or the length of the frieze that wraps around Trajan's Column.

NINE

My friend who liked reading track training books could hold his breath until he fainted. I just want to tell you that I am not asking this of you (the reader). Just hold your breath like a normal person, not like my friend from high school. So here is the exercise that I am giving all of you … yes, you, the one who is reading these words: when you read the word "Hollow," take a breath in and then hold it. Go ahead, I just said "hollow." I will wait … and you can now keep reading until you feel your body get to this point, to this pressure. Don't faint, now let your breath out. Take a few breaths. Then keep reading until you see the next placement of the word "hollow." This book is full of this word.

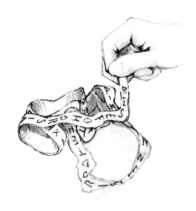

TEN

In fourteen steps we have traveled one section of the spiral. We have made one revolution, and now we are on top of where we were. Still we keep holding our breath. The view is overshadowed by our discomfort. Now we hold the pain and the light together. Step, step, hole in the wall then step....

ELEVEN

A scytale is a method for hiding messages, a way to securely pass a secret back and forth without fear of the enemy intercepting our note. The Romans in 500 BCE used this method. The system consisted of a piece of leather or parchment that was wrapped around a cylinder of a specific diameter. The leather spiraled around, then an important message was written horizontally across the length of the cylinder. Unless you had the same cylinder with the same diameter as the cylinder on which the message was written, you wouldn't be able to decipher it. The person who writes the note and the person who receives the note are the only ones with the correct cylinder. Imagine a larger cylinder something like 11 feet in diameter with a 625-foot strip wrapped around it.

TWELVE

Here the wooden cylinder (size of a wooden chair leg) is 1.25 inches in diameter. The half-inch strip of paper should be just under 6 feet long with this written on it: iolbahatfsdasevtistbcweetheeobdhbelnre altwdke deoiiwn btwtnop eoehgul ebroilr nuoutdm piftthi.

THIRTEEN

Wormhole court case: in the 1600s there was a case in which a worm was taken to court for eating a priest's chair. When he sat down he fell and the chair broke (presumably from the worm eating and not the weight of the godly man). Let us imagine that the worm entered into the wood after a long journey through the front door. It would have made it to the leg of the chair, eaten its way in and then chewed its way up towards the rest of the chair. We can imagine the moments when the worm eats too far to the edge of the cylinder of wood, the leg. This would create a window letting light into the tunnel the worm was digging. We can imagine this as a window allowing light in for the rest of the spiral up. But the hole might also have been to push out the chewings of the wood.

FOURTEEN

The walk makes us dizzy. Twenty-three revolutions, we have counted forty windows. We both exhale and breathe in. Our lungs fill. The air quickly passes back and forth from inside to outside. We prepare ourselves before we begin the walk down the hollow space.

Trajan's Hollow 1:10

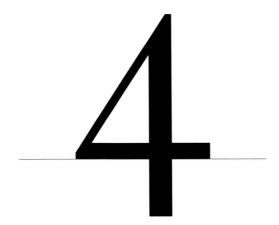

MARBLE IN THE 21ST-CENTURY

Trajan's Hollow 1:10 attempts to embody the language of marble-working into an object many times smaller than the original Column and the enormous quarry blocks of Carrara. To do so, this miniature reconstruction began by first reversing the process of the original's material journey, traveling backwards from Rome to the source of the marble in the quarries near modern-day Carrara.

Trajan, along with many of the ancient Roman builders, had a predilection for Luna marble, preferred because of the proximity of the quarry to the sea, which ensured easy access from Rome by boat.

Luna marble, along with the marble from nearby modern quarries of Carrara, is located in the Apuan Alps between Liguria and Tuscany. These mountains, formed during the Triassic, are an older, geological appendage to the Apennines. The famous stone here was developed approximately twenty-five million years ago when high pressure metamorphosed limestone of the Apuan Alps into marble.

While "Carrara" is usually used to describe the marble itself, there exists a network of specialized towns in the region, centered around the processing and transportation hub of Carrara. Much of the industrialized cutting operations associated with the marble used for construction and commercial purposes takes place in the city of Carrara, located between the mountains and the Ligurian Sea. The quarries themselves are located farther east in the mountains, while other towns in the alluvial plains such as Pietrasanta specialize in the sculptural marble-working craftsmanship of the *marmistas,* or marble carvers.

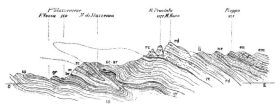

Sezione geologica attraverso il Monte Stazzema.

Paleozoico: *sp* scisti permiani. — **Trias**: *gr* grezzoni; *br* brecc.e; *m* marmi; *sc - ar* sc sti ed arenarie triassiche. — **Retico**: *rc* calcare dolomitico e cavernoso; *rd* dolomia superiore e portoro. — **Lias**: *li* calcari liassici. — **Neocomiano**: *ne* calcare bianco selcifero. — **Eocene**: *en* calcare nummulitico; *en* arenaria macigno.

DOMENICO ZACCAGNA, *GEOLOGICAL SECTION THROUGH MOUNT STAZZEMA*, 1920

THE EXPOSED MARBLE QUARRIES OF COLONNATA NEAR CARRARA (TOP)
AND THE BLOCK OF MARBLE SELECTED FOR *TRAJAN'S HOLLOW 1:10* (BOTTOM)

LOCAL AND GLOBAL EXPERTISE

Pietrasanta is the marble carving town near the quarries of Carrara. Since the
15th-century, when Michelangelo worked to access the quarries,
it has been the center of technical knowledge
for shaping the mineral wealth that emerges from deeper in the mountains.
This connection has recently been de-coupled
as the marmistas of Pietrasanta now sculpt with marble
imported from China and elsewhere in addition to stone from local quarries.

STAGETTI MARBLE STUDIO IN PIETRASANTA

HYBRID MARMISTAS: TRADITIONAL CRAFT VERSUS ROBOTIC LABOR

**The marmistas of Pietrasanta
have adapted contemporary technologies to marble sculpting.
The atelier of Marble Studio Stagetti
displays marble statuary sculpted by a team of skilled craftsmen
and a five-axis robotic milling machine. For most projects, the robotic mill
carves out the rough shape of the sculpture while the final carving and fine
detailing are completed by hand.**

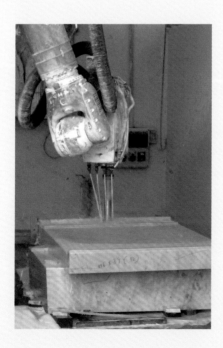

FIVE-AXIS CNC ROBOTIC MARBLE MILL AT STAGETTI MARBLE STUDIO

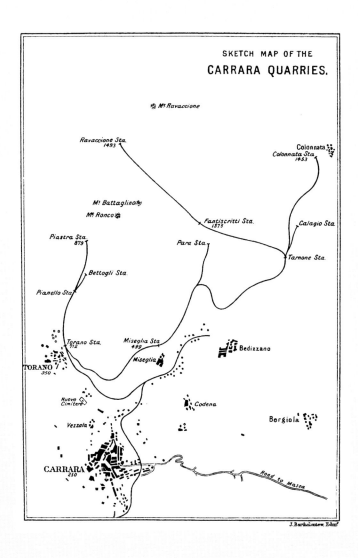

JOHN BARTHOLOMEW, *SKETCH MAP OF THE CARRARA QUARRIES*, 1896

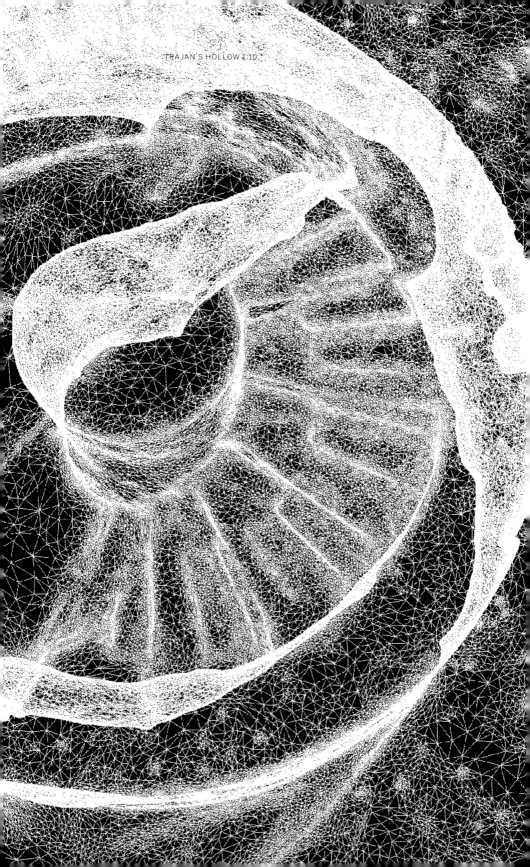

THE MARBLE DRUM IN MINIATURE

Scale = 10%

The miniature version of Trajan's Column—one-tenth its original size—retraces the journey of material from the Apennines while updating its construction history with contemporary techniques. This revised drum was designed to be fabricated out of marble from the modern quarries closest to the ancient Roman quarries of Luna where the marble for the Column was found. With this ancient material is introduced the exploration of new techniques: the 21st-century artisans in the marble carving town of Pietrasanta can now employ digital technologies to sculpt a miniature drum of the new Column.

This reconstruction of the drum was approached as an industrial design problem. While the original Column was constructed of nineteen unique cylindrical drums, in this revision, one drum is designed to be produced as identical copies stacked and rotated one on top of another. While the seam lines between the drums of the original are barely visible, now the asymmetrical drum opens each seam into a wedge of space to reveal its interior through these gaps.

This artifact embeds material and tooling patterns into its form. The diameter of the marble-carving robot's 5mm drill bit dictates the size and shape of the vertical cuts around the circumference of the drum. Each cut into the form indexes one figure in the narrative frieze. The drum was ultimately produced as a 3D powder print, which added a new layer of material grain and texture.

NARRATIVE FENESTRATION

The new apertures created by the CNC cuts (one per figure in the frieze) create a rhythm based on that of the exterior narrative. This series parallels and compliments the variation of existing window apertures, each connected to the specifics of the sculptural relief. Since the cuts represent one figure each, the rhythm of the narrative, if not the specifics, penetrates through to the interior.

From this new interior, the singular figure of Trajan stands out among the small groupings of his generals, or the larger mass of the entire army as it fords a river or encounters the Dacian forces. The ideal scholar of the frieze could "read" these cues and pinpoint their location within the new Column.

*Each bas-relief
figure generates a
tool path which
guides the CNC
carving tool along
a series of vertical
curves, which then
cut into the drum
according to the
frequency of figures.*

*Massing of
revised drum*

*Carving tool paths
matched to location
of each human
figure in
narrative frieze*

*Volume of material
to be removed by
CNC carving tool*

*Voids left by
CNC carving*

SEQUENCE OF OPERATIONS TRANSLATING NARRATIVE RHYTHM
OF BAS-RELIEF INTO APERTURE PATTERN

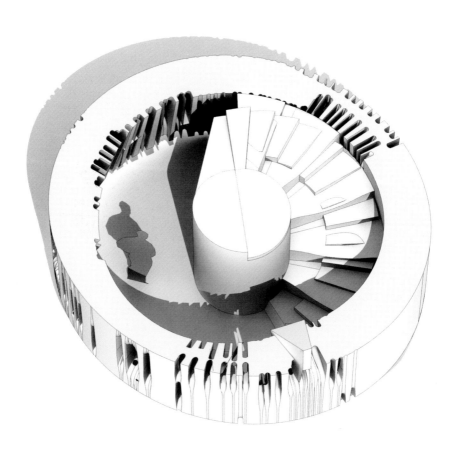

TRAJAN'S HOLLOW 1:10, VIEW FROM ABOVE

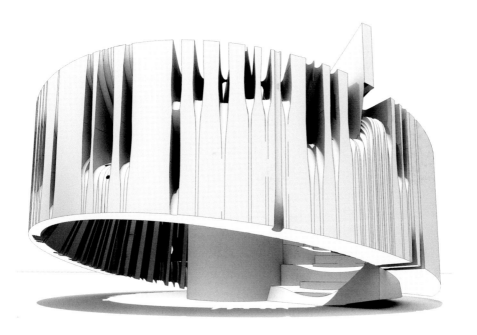

TRAJAN'S HOLLOW 1:10, EYE-LEVEL VIEW

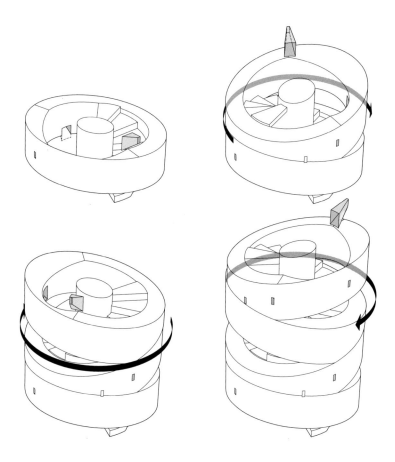

TRAJAN'S HOLLOW 1:10, STACKING LOGIC

*As they stack,
each drum rotates
206.55° creating
a new aperture
or gap between
its neighbors above
and below.*

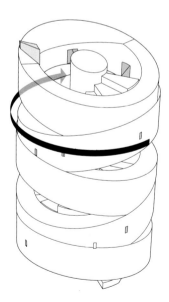

PROJECTIVE FIGURATION

While each figure in the bas-relief frieze protrudes out by about 3cm,
in *Trajan's Hollow 1:10* each figure within the narrative frieze projects inward
by about 15cm, carving out space. A mirrored version of the frieze from the
inside means that when these projections meet, new apertures are created
according to the frequency of activity in the narrative.

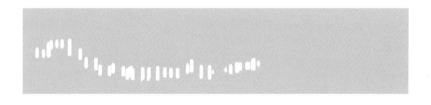

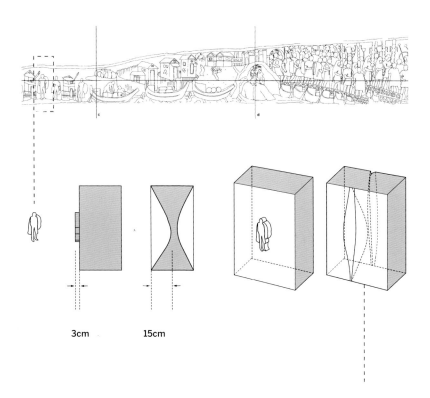

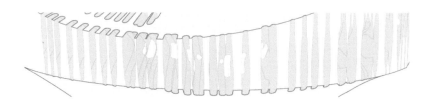

DIAGRAM OF TRANSLATION OF FIGURAL BAS-RELIEF INTO APERTURE CUTS

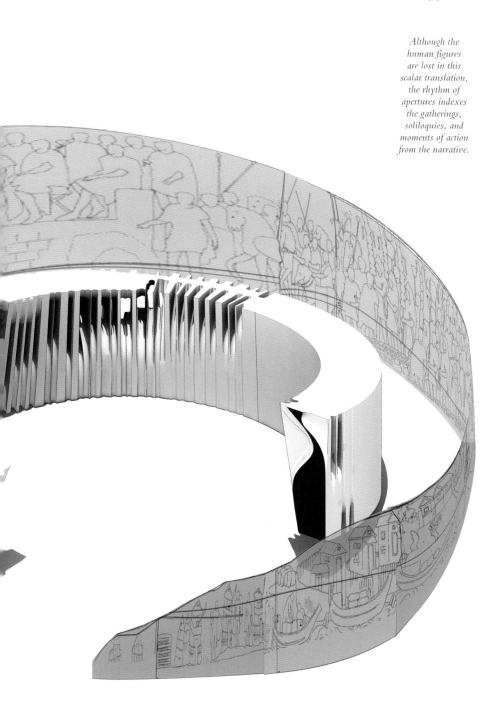

*Although the
human figures
are lost in this
scalar translation,
the rhythm of
apertures indexes
the gatherings,
soliloquies, and
moments of action
from the narrative.*

RELATIONSHIP BETWEEN NARRATIVE FRIEZE AND PROJECTED APERTURES

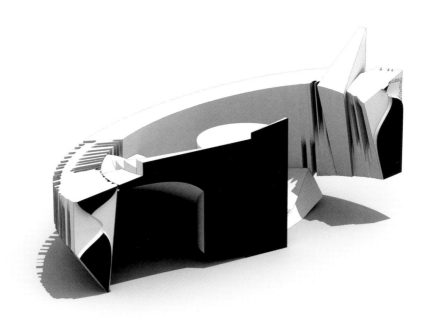

VERTICAL SECTIONAL MODEL OF *TRAJAN'S HOLLOW 1:10*

The horizontal cross-section forecasts the form of the Trajan's Hollow 1.5 floor cast.

HORIZONTAL SECTIONAL MODEL OF *TRAJAN'S HOLLOW 1:10*

Trajan's Hollow 3D-Printed Model

TRAJAN'S HOLLOW 1:10

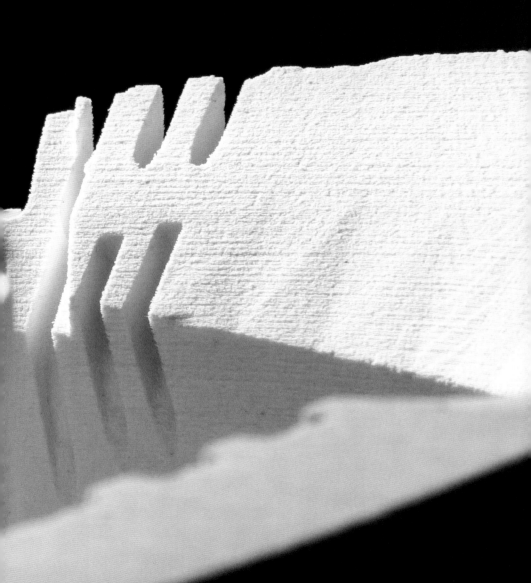

Scaling Trajan's Hollow

The Collas Machine invented by Achille Collas in the 19th-century could simultaneously trace and carve sculptural replicas.

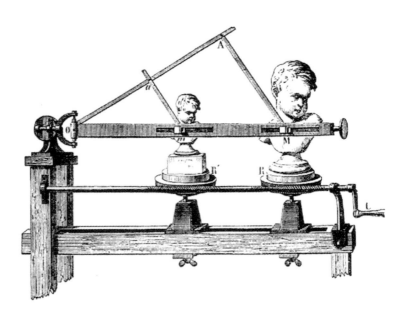

ACHILLE COLLAS'S 1837 DRAWING FOR MACHINE USED TO REPRODUCE
AND ENLARGE MAQUETTES

The Agency of Scale:
Employing Scalar Shift as a Design Strategy

JOSHUA G. STEIN

The ability to enlarge form took a momentous leap in the digital era. A host of software modeling packages gained popularity in the 1990s, fundamentally altering architecture's relationship to scale and object-hood. The virtual modeling space of digital software, removed from the material constraints of the physical world, has effectively created an a-scalar environment in which dimensional specificity is completely fluid and temporarily fictional. This digital scalelessness offers the potent capability to move effortlessly from the small to the big and back again—as long as one remains in this virtual space. However, this newfound facility within this a-scalar design environment has also eroded an important disciplinary expertise in scale and its effects in architectural design. Current advances in technology demand a reconsideration of these techniques and their design potential.

Throughout history artists, craftspeople, and designers have continually been compelled to make things bigger—or, more precisely, to produce copies of objects they find in the world or create themselves at a scale different from the original artifact. A clever artisan might make a small-scale version of a proposed project to visualize form or composition before expending the labor and materials necessary to produce the "real thing." But not all enlargements are created equally. The primary challenge is to develop amplification techniques that maintain greater precision than pictorial approximation or gestural "eyeballing." Contemporary techniques of digital scanning and reproduction now allow us to move beyond a simple visual enlargement and instead integrate the less predictable influences of texture, materiality, and tooling into the process.

The history of inventions aimed at reproducing two-dimensional sketches and three-dimensional maquettes into much larger tableaus and sculptures stretches back at least 5,000 years. Squaring, employed by the Egyptians, involved the use of a grid mapped out on a smaller object to then guide the transfer of shapes to the same grid expanded to a larger scale. More mechanized methods would follow starting in the Renaissance. With the pantograph, invented in 1603, artists could

The pointing device was developed in the 18th-century to guide in the one-to-one reproduction of three-dimensional sculptural form.

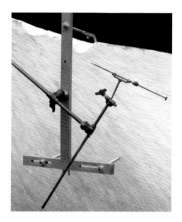

The pantograph was a 17th-century invention used for two-dimensional reproduction, reduction, and enlargement.

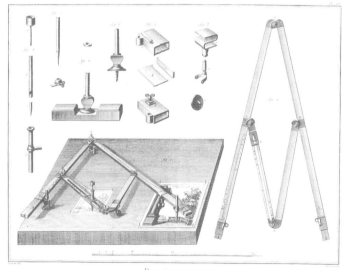

POINTING DEVICE (TOP) AND ENGRAVING OF A PANTOGRAPH (BOTTOM)

FROM DIDEROT ET D'ALEMBERT'S *ENCYCLOPÉDIE*, 1751–1772

The Collas Machine was used by Rodin's fabricators to replicate, reduce, or enlarge his maquettes.

AUGUSTE RODIN'S *THE THREE SHADES*, FIRST VERSION (TOP),
MODELED 1880–1884, BRONZE, MUSÉE RODIN CAST 10 IN 1981,
38 1/4 × 37 1/2 × 20 1/2 IN., COUBERTIN FOUNDRY (IRIS CANTOR COLLECTION),
AND ENLARGED VERSION (BOTTOM), 1898, BRONZE, 75 1/2 × 73 1/2 × 41 IN.

proportionally trace an image with a stylus while drawing a larger one with a pen held by four simple straight lengths properly pinned to one another. This machine suppressed the human margin of error but amplified the challenge of mechanical precision and friction—visual misinterpretation by the artisan became less an issue while properly sharpened tools and well-oiled joints became more critical.

For three-dimensional work, the task was more difficult. The pointing machine invented in the late 18th-century allowed a sculptor to proportionally transfer a set of dimensions from a maquette to a larger block of marble or other material. This simple strategy of indexing coordinates was a three-dimensional corollary to the squaring method and is similar to the digitizer arms used today. It was an improvement over earlier techniques, but still demanded that the artist interpolate a great deal of information by eye, since the device could only point to where sculpting was necessary, but not itself sculpt. In 1836, Achille Collas invented the Collas Machine, a rotational device that worked much like an industrial replicating lathe, indexing the form of a small object while carving its likeness just adjacent. This would prove to be equivalent to a pantograph for three-dimensional applications. The Collas Machine was used by Henri Lebossé, Rodin's primary fabricator,

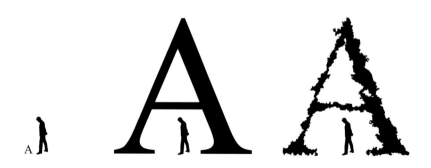

to produce the many replicas at multiple scales that Rodin's busy atelier demanded. While a series of advances followed, from mechanical to optical to digital, which drastically increased the speed of two-dimensional scaling, three-dimensional scaling would only recently be reinvented by the proliferation of digital scanning and fabrication.

With the advent of vector illustrations and NURBS modeling, it seems easy enough to output from the digital world to the physical at any scale. In this fluid, immaterial landscape, pixel files and analog processes are relegated to inferior status as formats for scaling. Although a vector file of the letter "A" might infinitely scale up without ever succumbing to degradation, that same two-dimensional form enlarged on an optical photocopier inevitably becomes increasingly distorted. However, instead of a loss of quality, this transformation could be understood as the emergence of physical behaviors over semiotic readings. While the scalelessness of the virtual allows for the easy projection of form beyond the confines of the immediate, it does so at the expense of material specifics inherent in all analog processes.

THE ARCHITECTURAL SURROGATE AND FRICTIONLESS SCALING

Architecture has always relied on both two-dimensional and three-dimensional miniature surrogates for representing design proposals and effecting them into full-scale architectural projects, what could be called "built works." Architectural models have existed for millennia: we can find examples from as early as 4600–3900 BCE from the Gumelnita civilization in the Danube Valley. These models were not necessarily used to represent a project for construction, but instead as gifts or in burial ceremonies. However, models from the early Renaissance were used more in the manner that we use them today, to visualize and develop massing, detailing, appearance, and structure. Ratio and proportion, so fundamental in classical and Renaissance aesthetics, were also utilized to translate designs from model scale to that of built work.

While architects regularly engage in scalar translations, this is most typically in the service of representation: the architectural model is used to represent some projected version of a future built work. In general terms, we can differentiate between size, the absolute dimensions of an object, and scale, the relative ratio between one object and another of different dimensions. But perhaps more important in this discussion is the action of moving from one scale to another. Typically, this involves the proportional increase of all the dimensional features of an object according to a set ratio.

The advent of the digital or virtual model demands a reconsideration of how we conceive of scale, since the lack of physicality in digital modeling denies any verifiable way of tracking scale. In most software packages, one models either without a particular scale, or at "full-scale" where the singular set of dimensions assigned to a digital form simply anticipates the dimensions of the projected built work. This functions differently from physical models, which would be constructed with two sets of dimensions: those of the model itself as well as those of the proposed built work, regulated by the proper ratio that relates one set of dimensions to the other (ie. 1:100). In the space of virtual modeling, the fact that there is only one set of dimensions, that of the proposed built work, means that there is no corresponding ratio and therefore no scale—one can "resize" any digital model without any real repercussions. The development of physics simulators is slowly bringing more meaning to the notion of virtual scale, but for the moment, scale is directly yoked to physicality, gravity, and materiality.

We know that the materials employed in physical model-making have a significant influence on both the formal and representational effects of a model. Any architecture student understands that constructing a model from linear elements (basswood sticks), versus planar (chipboard), versus volumetric methods (cut foam or cast plaster) will each offer a very different means of rendering form and tectonics. Most architectural models attempt to neuter the power of their materials so that they can better project to some other scale and physicality. Basswood or foam are often selected precisely for their lack of qualities like grain. These models are solely interested in representing the proposed built work and ask nothing of the substantive physicality that exists at the scale of the surrogate model.[1] They are then scaled using geometric and proportional relationships that preserve the visual appearance instead of following material allegiances. This dichotomy aligns with Didi-Huberman's distinction (see page 133) between the imitation which inhabits the regime of the visual (primarily concerned with immaterial properties like profile and geometry) and the imprint of the material regime (concerned with more tangible qualities).

However, a quick look into history reveals an alternative approach to the relationship between matter and form. The form-finding models of Antoni Gaudí or Frei Otto operate as physical computational machines where there is no preconceived geometry, only a precise set of procedural rules that are set into literal motion through material interactions. For these form-finding machines to operate successfully, they must be understood as existing at their own independent scale, free to move without being tied to the role of representing some other scale to come. This craft-like immediacy and primacy of the physical artifact allows for the architect to explore without a priori notions of representational form. The term "live models," coined by Jason Kelly Johnson and Nataly Gattegno, describes models that "can be used as analytical engines to understand the patterns around us, and in some cases, as conceptual frameworks for architecture."[2] These constructs are resolutely of their own scale and substance, while simultaneously pointing to a potential built work at some larger scale and unknown material. They are both simulations and representations. Marginalized under a modernist notion of ideal form and geometry, these forays into craft sensibility and scalar immediacy have recently been reintroduced into the discourse of the discipline. In the last two decades a number of architects have picked up the mantle of experimenting with live models. Lars Spuybroek / NOX has worked on numerous projects that explicitly seek to advance the experiments initiated by Gaudí and Otto which, while time-consuming and labor-intensive, may find new relevance as emerging technologies allow architects to practice contemporary methods of form-finding and iterative scaling.[3]

Although there are clear difficulties in re-producing form at the scale of a building based on the behaviors of a miniature surrogate, Gaudí and Otto focused their research on the aspects of physics and structures that are capable of operating across multiple scales. The results are powerful precisely because their shapes remain constant as they move from one scale to the next. However, not all material properties scale easily; most importantly, the surface to volume ratio does not remain constant across different scales which means that structural solutions at one scale will not necessarily hold true at another. While Otto's minimal surface studies and Gaudí's largely two-dimensional experiments with catenary vectors prove that it is possible to maintain almost identical configuration across scale shift from live model to architecture, they rarely attempt to deal with the issues of volume or mass.[4] Following these innovative studies would limit all form-finding exercises to minimal surface and catenary studies, reproducing the same predilections towards scalar translations that focus on reproducing visual profile, most

Trajan's Hollow 1:10

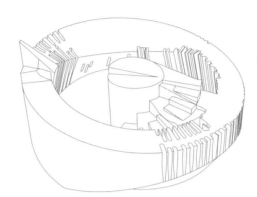

Slice from 1:10
model used for
Trajan's Hollow 1.5

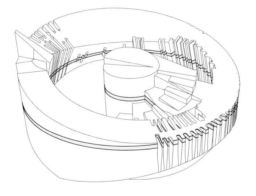

Trajan's Hollow 1.5
floor cast

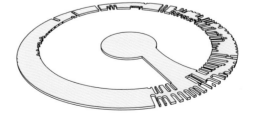

RELATIONSHIP BETWEEN *TRAJAN'S HOLLOW 1:10* AND *TRAJAN'S HOLLOW 1.5*

easily accomplished through simple proportional scaling. However, their research lays a crucial foundation for examining the computational power of materials as they interact with friction across multiple scales.

THE GENERATIVE FRICTION OF SCALE SHIFT

It is important to examine not only the techniques that allow for fluid translation from one scale to another (most often visual, geometric, and numerical), but also those that create friction and produce disruptions in this process (most often material). If we believe that matter has agency within the design process and is not simply the vessel for content or form, then it would seem a worthwhile endeavor to explore its contribution to the process of scale shift. While much of architecture has focused on frictionless scale shift, other disciplines have also investigated an oppositional phenomenon that I will call "frictious" scaling—a version of scaling where material constraints disrupt easy proportional scaling. In the biological world, the process of differential or nonproportional growth across a shift in scale follows clear morphological constraints. Allometry is the branch of biology specifically concerned with this non-isometric growth. D'Arcy Wentworth Thompson's book *On Growth and Form*, best known to architects for its research into the impact of physics and mechanics on the evolution of biological organisms, is also one of the foundational texts on allometry. Thompson's colleague and friend J. S. Haldane, a scientist who has written at length on allometry, points out that the relationship of surface area to mass and volume means that biological forms that thrive at one scale do not necessarily function at another: "For every type of animal there is a most convenient size, and a large change in size inevitably carries with it a change of form." If matter and the forces which operate on it are taken into account, form registers these effects and alters as it scales up or down. We might refer to this as the friction associated with scale shift or frictious scaling.

This friction is not literal. It is less about the movement of two surfaces against one another and instead more of a non-quantitative coefficient, a qualifier tied to the aberrations produced by the dictates of the physical realm during the shift from one scale to another. As in other domains, friction can either be perceived as a disruptive force that destroys predictable calculations, or as a generative force that not only firmly tethers abstract calculations to the world of material phenomena, but produces an increase in complexity, specialization, and speciation. As in the case of enlarging the letter "A," the easy operation of frictionless digital scaling produces no new complexity—the "A" is the same, only larger. However, in enlargement by analog means, the friction from scaling inflects the "A" with new crenellations and complexities.

At different scales organisms must respond to different "regimes of force": while animals larger than a mouse are most affected by gravity, smaller animals are scarcely concerned with falling but are instead governed by the effects of surface tension—imagine a water strider. One of the most basic principles of allometry is that as an object or organism scales upwards, its "body plan," the formal and organizational diagram of an organism, may have to adapt to the different forces operating at this new scale. In moving from the very small to the very large, the transformation of body plans manifests a shift in allegiances to different force regimes: certain scales are governed by conditions that produce dispersed patterns without any hierarchy, while at other scales, clear figuration emerges. The defining stills from the film *Powers of Ten* by Charles and Ray Eames (1977) demonstrate an oscillation between the figural (Earth, Lake Michigan, picnic blanket) to patterns both centralized (Milky Way, convergence of crevices in human skin) and field-like (atmospheric patterns, city grids).

As organisms and phenomena are pushed across differing scales we can imagine that certain traits emerge and retreat, responding to the pressures of each new scalar context. These sets of different traits could exist simultaneously: those no longer useful at one scale might simply become recessive, only to later reappear at another scale. This simultaneity of different traits aligns with the increased number of organ and tissue systems found in larger organisms in comparison to smaller ones. Haldane explains this necessity in "On Being the Right Size":

> While vertebrates carry the oxygen from the gills or lungs all over the body in the blood, insects take air directly to every part of their body by tiny blind tubes called tracheae which open to the surface at many different points. … In consequence hardly any insects are much more than half an inch thick. … If the insects had hit on a plan for driving air through their tissues instead of letting it soak in, they might well have become as large as lobsters, though other considerations would have prevented them from becoming as large as man.

Therefore, as organisms increase in size, they also complexify through the multiplication of different internal systems. In Haldane's example, because an insect effectively breathes through its skin, it does not need a circulatory system to move oxygen to different parts of its body. Larger organisms would require respiratory and circulatory systems to accomplish the same task. We can see this as a division of labor that produces a layering of distinct and specialized systems within larger entities, either biological or mechanical. This diversification of systems offers the designer a curatorial choice as to what traits remain most dominant as form is moved from one scale to the next. However, not all of these traits can exist in equal amounts simultaneously. In fact they are often in competition with one another, introducing the dilemma of competing fidelities.

COMPETING FIDELITIES

A parallel series of projects from Claes Oldenburg tests different responses to the competing notions of fidelity demanded in scale enlargement. The giant sculptures produced by Oldenburg and collaborator Coosje Van Bruggen loom large in their career not just because of their magnitude, but because they are recognizably over-sized versions of objects from a smaller scale. At first, these giant sculptures are so powerful in the urban and gallery settings precisely because of their seemingly emphatic denial of the effects of scale shift on material behavior: the household flashlight looks exactly like the one we know so well, but has been defamiliarized by its gargantuan size. However, examining the large works reveals a range of propositions regarding scaling that are more nuanced than that of the public perception of their giant hard sculptures.

In Oldenburg's *Light Switches – Hard Version* of 1964, we see a typical double-switch plate enlarged to over ten times its typical dimensions. Its proportions are nearly identical to those of the light switches we might find at any hardware store. Although the piece is a convincing visual facsimile of mass-produced switches and plate, if we could glimpse the back side of this oversized object, we would find it is crafted of painted wood and metal: a radically different tectonic from the molded thermoset plastic we would expect. While the gargantuan switch aims for fidelity to the figural profile of its smaller progenitor, it abandons any expectations of faithfulness to its physical reality. By contrast, Oldenburg's *Soft Light Switches – "Ghost Version" II* tests a radically different set of allegiances. While this version starts by approximating the sharp edges of the original form, it is constructed of gesso over sewn canvas stuffed with kapok (a mattress filling product). The material does not have the strength to maintain the crisp profiles expected of it and instead buckles and sags under its own weight. But while it uncomfortably lacks the hard version's precise visual likeness to a typical light switch, it demonstrates a more accurate expression of matter's relationship to its surrounding forces as it reacts to a drastic shift in scale. In addition, it unabashedly displays the means and materials of the craft involved in its making.

Between these two instantiations we can see Oldenburg testing different loyalties: first to visual profile and then to material behavior and craft. How would the light switches appear if their form followed other fidelities? Focus on the techniques of digital scanning might produce something more like a diaphanous point cloud, while an allegiance to the division of labor could offer clear demarcations between each phase and trade involved in production. Confronting the perspectival effects of enlargement would result in something like the entasis employed in Greek columns or Michelangelo's response to foreshortening in his David, while adherence to material affordances might fashion a mosaic of different substances and tectonic patterns where each is used only in the context where it best responds to the challenges of curvature, detail, and massiveness. Although it may be outside of his explicit intentions, Oldenburg's enlargements fracture any hope of reproducing and scaling a form along with all of its attributes intact. Instead we must recognize a delamination of fidelities, which the artist prioritizes and orchestrates.

Historically, in the world of sculpture, the primary concerns in enlargement would be profile, proportion, and texture. Certain techniques of enlargement offer more precision or fidelity to these traits than others. While the Collas Machine could render profile and proportion at a different scale with relatively fine precision, it could never reproduce the level of textural detail that a plaster cast could. However, casting techniques offer relatively little help in the challenge of scaling. Digital scanning aids immensely with moving across scale but like plaster casting, captures a high degree of surface detail while struggling to resolve coherent profiles.

As it is physically impossible to maintain infinite fidelity across all domains during enlargement, the designer must prioritize among different possible affiliations. The dream of an ultimate fidelity through analog reproduction—epitomized by "hi-fi" stereo fantasies of the

Although Oldenburg's giant hard sculpture denies any effects of scale shift, his giant soft sculpture acknowledges the weight of materiality and displays the seams of its sewn construction.

 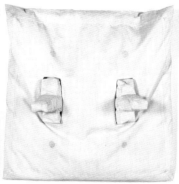

CLAES OLDENBURG, *LIGHT SWITCHES — HARD VERSION*, 1964, PAINTED WOOD AND METAL, 47 3/4 × 47 3/4 × 11 3/4 IN. (RIGHT) AND *SOFT LIGHT SWITCHES — "GHOST VERSION" II*, 1964–71, CANVAS FILLED WITH KAPOK, GESSO, PENCIL, 47 × 47 × 12 IN. (LEFT)

1950s—hoped to produce a copy completely free of any friction, static, or marker of its medium. Reignited during the digital revolution, this dream reemerges as a mirage of infinite precision and identical likeness hovering ever closer on the horizon. And while there may one day be the possibility of infinite fidelity within the realm of one-to-one facsimiles (however much this cuts against the grain of material realities), to achieve this coupled with scale shift will always prove a physical impossibility. As long as gravity is involved, Moore's Law is no match for Newton's.

Instead, scale shift splinters the replication of form into a Pandora's box of challenges, pitting competing desires for fidelity in one domain against those in another. Most architects understand that they cannot follow both the form in a physical model and the same construction techniques. Locating or producing materials in panel sizes that correspond to direct enlargements of the pieces of basswood in a model would be a folly. Therefore, the architect needs to choose between either the model form, subdividing large surfaces into smaller components, or the tectonic reality which might propose an aberrant version of the surrogate model. Of course it is usually form that wins in this contest. It is no coincidence that panelization innovation in façade engineering is one of the primary applications of digital design at the moment. This contemporary agenda levies an immense amount of academic and corporate resources armed with the latest computational tools to determine ever more sophisticated means of breaking down a building façade, while leaving uninterrogated the more fundamental questions of how scaling might affect building type or configuration.

The unbundling of fidelities is not necessarily something to lament. This delamination and multiplication of different possible fidelities and the impossibility of achieving all simultaneously, underscores the strategic and curatorial power of the designer. While we assume that larger organisms' biological systems all work in concert towards a unified goal, this collaboration is the result of millennia of evolutionary development. In the world of design, where scale shift is often a quick and violent process, the multiple systems necessary in a large object may end up working at cross purposes, or simply ignorant of their associated systems. The architect has the ability to draw out different qualities that either align with or stand against the expected "comfortable" solutions. These very issues should be strategically employed within architectural design, allowing the process of scaling to create productive relationships between differing fidelities.

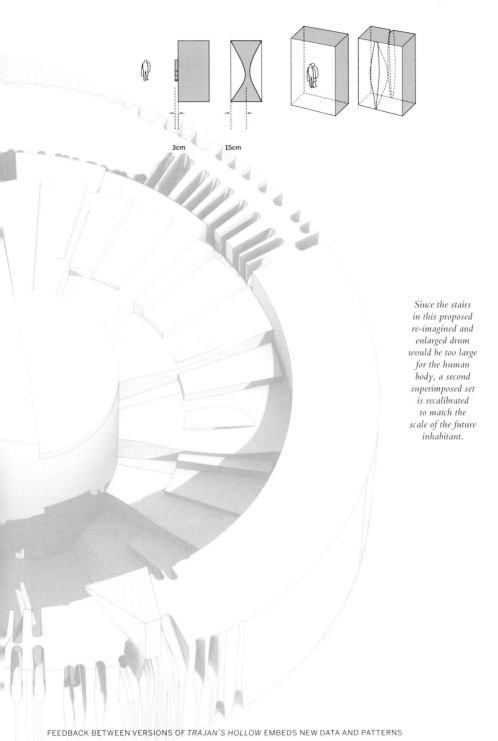

3cm   15cm

Since the stairs in this proposed re-imagined and enlarged drum would be too large for the human body, a second superimposed set is recalibrated to match the scale of the future inhabitant.

FEEDBACK BETWEEN VERSIONS OF *TRAJAN'S HOLLOW* EMBEDS NEW DATA AND PATTERNS INTO EACH DIFFERENT SCALE ITERATION

SCALING AS READING AND WRITING: TRAJAN'S HOLLOW

Trajan's Hollow explicitly employs scalar translation—in both directions—to radically alter the form of the original, ultimately transforming our perception of its qualities and effects. The project was conceived as a set of iterative reconstructions of the original Column: a miniature version of one of the monolithic marble drums at one-tenth the size of the original and a slice of the same drum at 1.5 times the size of the original. With the original Column included, this produced a set of three cross-indexed versions at 100%, 10%, and 150%. Each instantiation of *Trajan's Hollow* is approached as a one-to-one exercise that is linked to iterations at other scales. Modifications to one instance of *Trajan's Hollow* induce recalibrations that ripple across all other editions. The dimensions for the window apertures carved into the 1:10 scale drum were determined by the dimensions of the six-millimeter diameter marble-carving tool used for this small piece. At this scale, one-tenth of the original monument, the available techniques could never have faithfully sculpted the bas-relief human figures with any degree of legibility. In fact, the specific scale of this 1:10 "model" was selected to maximize the impact of these reductive operations as apertures instead of ornament, anticipating their relationship to the human experience in the 150% enlargement. These same incisions appear in this larger reconstruction, but their dimensions are now no longer tied to the tools of fabrication. They have become vestigial artifacts of frictious scaling that can then be intentionally re-appropriated for another purpose. At the larger scale, a new set of tool marks is layered atop these cuts that originate in the smaller version: crenellations from hand-cut foam mold parts, casting seam lines, chisel marks in plaster—all update and complicate the effects of tooling from another scale.

    Trajan's Hollow exercises frictious scale shift to realign allegiances of reproduction in order to escape the figural representation of narrative as embodied in the monument's obsessively documented frieze and instead prioritize the phenomenological experience of the inhabited Column. The scale of 1:10 offers a reprieve from the historiographical tendency to prioritize the visual legibility of the bas-relief sculpture and instead offers the architectural potential to experience the narrative rhythm tactilely at the 1:10 scale and bodily, at the 150% size through the haptic encounter with space, light, and aperture. *Trajan's Hollow* intentionally slides between different scales accumulating and shedding certain allegiances as its form lodges itself in various materials and sizes.

By layering in new patterns, formal and experiential, generated by frictious scaling, *Trajan's Hollow* also embodies a simultaneity of experience. The reconstructions fuse two sets of stairs, Apollodorus's eroded
originals scaled up by 150% plus a second set superimposed on top.
Since the stairs in the re-imagined and enlarged drum would be too
large for the human body, the second set is recalibrated to match the
scale of the human body. The fact that these two staircases intersect in
the same space allows the visitor to recognize that a scale shift has occurred and therefore conceptually occupy two moments in history and
space simultaneously (the original Column of 113 CE in Trajan's Forum,
and the 2011 reconstruction at the American Academy in Rome). In
Trajan's Column, each tread of the spiral stair of the Column has been
inadvertently sculpted into gentle miniature landscapes by thousands
of years of human and environmental weathering: the friction from
footsteps and penetrating rainwater have worked in tandem to transform the steps into a series of cascading basins (see page 62). The new
inserted stairs are "reset" back to their second century dimensions and
shape. When set against their transformed original, as one moves from
step to step, one inhabits the effects of both time and scale shift as they
rework the marble stone (see page 213). The discrepancy between the
two sets of stairs makes evident the almost geological transformation
of the marble over 2,000 years. As in many archeological dig sites, this
also allows the public to follow on a path parallel to the original yet one
step removed—somewhere between retracing the steps of the ancients
at eye level and analyzing their movements from above. This echoes the
walkways of steel grating that hover over so many Roman archaeological sites across the former empire (or closer to the Column's home, the
new COR-TEN steel footpaths inserted into Trajan's Market by Labics
Architects in 2004).

Trajan's Hollow amplifies the simultaneity of reproduction as both
documentation and transformative reinvention. The pantograph or
Collas Machine physically links an object to its copy. Usually one side
of the machine reads and one side writes—at one of the nodes is a stylus,
meant for tracing form, while the other is a drill bit which carves into
the marble. Each of the two nodes can accept either reading instrument or writing instrument and, when reversed, produces a machine
which can toggle from enlarging to reducing dimensions. *Trajan's Hollow*
conceptually operates as a Collas Machine in which each tool can both
read and write simultaneously. With the exception of the original
Column, which was of course impossible to modify physically, each

instantiation of *Trajan's Hollow* "reads" information from the set of prior artifacts, instantaneously "writing" data back into this family of fraternal twins—a simultaneity of relayed patterns between networked siblings where each new layer of pattern or parameter becomes a sedimentary deposit integrated into each copy. The causal relationship between these reproductions remains fluid, in keeping with Sigfried Giedion's point that "History cannot be touched without changing it." This supports the tenets of the observer effect, in which the instruments of scientific measurement by their very nature alter the phenomenon in question.

After decades of working with digital pixel-based image files, we now assume that information is added during enlargement and deleted during reduction. While much of my earlier discussion focuses on bio-logical analogy to track transformation of form through the process of enlargement, *Trajan's Hollow* demonstrates that both amplification and miniaturization can add complexity into the scaled artifact through frictious scaling. Regardless of the direction of scaling, the act of scalar translation itself increases the information available to the designer. In *Trajan's Hollow 1:10*, the tooling data from the CNC router is added to the existing bas-relief. Of course this new information could also be discarded, which might fall into a more conventional architectural pro-cess, but it is precisely this default tendency that this project attempts to challenge.

STRADDLING SCALE

Ceramists often employ a device called a "shrink ruler," which has two sets of dimensions, one tied to the size of wet or plastic clay and one tied to the ultimate dimensions of the final piece after firing. As differ-ent clay bodies have different shrink rates (typically between 5% and 15%), a ceramist may have multiple rulers, one calibrated for each type of clay. The shrink ruler helps the ceramist work at two different scales simultaneously. Historically, physical modeling and drafting forced the architect to engage in a similar simultaneity of scale. The architect's scale, a physical tool that converts the dimensions of an architectural model to those of the proposed built work, plays a role similar to that of the shrink ruler, connecting much more extreme differences in scale. Until the 1990s, this guaranteed that the architect would have a leg in both scales at once, but current digital modeling practices have undermined the practice of "inhabiting" two worlds simultaneously. Although architecture is one of the few disciplines tasked with reconciling constraints across multiple scales, we spend surprisingly little time discussing the charged relationship between architecture as scalar surrogate and architecture as built work. The fact that a profession so entangled in problems of scale has produced so little theory addressing this issue reveals a persistent lurking disinterest in the effects material and friction exercise over form and inhabitation.

However, while digital modeling can be accused of undermining the architect's control of scale shift, it is this same shift in technology that offers architects a powerful yet underutilized design tool through scaling. Virtual modeling space provides a quick way to momentarily operate outside of scalar constraints before re-introducing a schema back into a specific tangible scale, accumulating material data upon re-entry. This facility offers an important relief from the time-consuming, intensive methods of mechanical scaling like the Collas Machine—similar to disengaging the clutch while shifting gears in a manual transmission car. While this momentary lapse of responsibility to the physical realm speeds the workflow, it does not produce new schemas or body plans. This is what frictious scaling offers, depositing new strata of material information and competing fidelities.

LAYER MANAGEMENT

The capabilities of digital applications map directly onto the challenge of orchestrating these multiple layers of data and potential allegiances. The cognitive process of layer management can negotiate the simultaneity of different sets of information, each of which correspond to one particular material scale. Software can "compress" or archive recessive qualities that may not have immediate relevancy at one scale but may be "unzipped" to more potential at another scale. Most digital technologies associated with scanning, modeling, simulation, and fabrication already operate through a logic of multiple layers, passes, and particles that parse large ambitions into smaller discrete entities. Laser scanning traces three-dimensional form and texture as a stacked series of two-dimensional paths or as a cloud of points. Three-dimensional printing assembles particles together, often layer by layer. CNC carving similarly cuts paths that successively add up to something that describes a surface.[5] Each of these processes leaves a trace which may or may not be evident in the material artifact. Archiving and managing these different layers of data means the architect controls their delamination and reorchestration, which can then add new meaning to, support or contest patterns of fidelities at other scales.

This would require that architects move beyond the current conception of scale as either a linear set of workflow relationships or a static set of contextual constraints. In the former, scale implies a sequence of increasing levels of development or resolution in executing a built work: the architect first considers urban implications, then sketches out massing, form, and internal programmatic relationships, then structure and mechanical, then finally perfects detailing at the scale of tectonics. In the latter, scale indicates a nested set of various constraints that dictate design decisions through each of these phases: environmental constraints at a regional scale, building code constraints, and anthropometrics at smaller scales. What if scale were not just a sequence or context within which architecture plays itself out, but rather an operation that reinvents these constraints? We could imagine scaling as a generative series of nonlinear and dynamic translations that simultaneously reconfigure and reprogram existing schemas.

In the larger oeuvre of Oldenburg and Van Bruggen, scaling is often employed as a productive alibi though which to introduce other aberrations beyond simple dimensional enlargement—formal abstractions, humorous associations, and visual modifications augment or undermine the initial reading of an enlarged object. However, in the *Soft Light Switches – "Ghost Version" II*, frictious scaling not only offers the excuse, but also the means to refigure and reprogram the light switch. The resulting transformations are more than the imposition of an idiosyncratic aesthetic on the part of the artist. Instead, frictious scaling relaxes the artist's formal control while increasing options for artistic curation, materializing forms that emerge from allegiances guided by a selective fidelity. This opens new possible interpretations that transgress the semiotic reading of a giant light switch: this no longer looks like a light switch and no longer has to be compared to one. The emergent slumped form reconfigures the light switch formally, almost, but not quite, beyond recognition. In addition, frictious scaling "reprograms" the initial artifact, forcing us to now ask "What would be the affect and performance of regular light switches if they were floppy?" If, like *Trajan's Hollow*, there is still open feedback between multiple scales, frictious scaling can force us to rethink both source object and its scaled cousin.

Frictious scaling might then be used even when enlargement or miniaturization is not required, but instead as a way to interrogate the design
or schema at hand. Similar to Haldane's description of new body plans
that emerge as organisms move across scale, we might imagine frictious
scaling as generating new plans, schemas, partis, or architectural diagrams,
refiguring and reprogramming both the original and its kin at differing
scales. The task for the discipline is to move beyond trafficking in *scale*,
and to instead explore *scaling*.

Portions of this essay were developed in collaboration with Glenn Adamson for the
article "Imprints: Scale Shifts and the Maker's Trace" by Glenn Adamson and Joshua G.
Stein appearing in the volume *Scale* edited by Jennifer L. Roberts (Terra Foundation for
American Art, 2016).

1. See Joshua G. Stein, "Speculative Artisanry: The Expanding Scale of Craft within
   Architecture," *The Journal of Modern Craft,* v.4, no. 1 (March 2011): 49–63.

2. Jason Kelly Johnson and Nataly Gattegno, "Live Models" in *Dirt*, ed. Megan Born,
   Helene Furján and Lily Jencks, (Philadelphia: PennDesign, 2012), 143.

3. See Lars Spuybroek, "Structure of Vagueness," in *NOX: machining architecture*, (London:
   Thames & Hudson, 2004), 352–359.

4. Otto has developed investigations into the angle of repose of aggregates, but these studies
   have found few architectural applications.

5. For more see Joshua G. Stein, "Production Perforation: A New Taxonomy of Texture," in
   New Morphologies: Studio Ceramics and Digital Processes, ed. Del Harrow and Stacy Jo Scott,
   (Alfred, NY: School of Art and Design, New York State College of Ceramics at Alfred
   University, 2014), 26–31.

AMERICAN ACADEMY IN ROME STUDIO SPACE WITH
FULL-SCALE FOOTPRINT OF TRAJAN'S COLUMN

AMERICAN ACADEMY IN ROME STUDIO SPACE WITH

FOOTPRINT OF *TRAJAN'S HOLLOW 1:10*

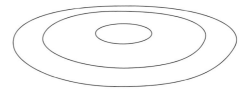

FULL-SCALE FOOTPRINT OF TRAJAN'S COLUMN

AMERICAN ACADEMY IN ROME STUDIO SPACE WITH
FOOTPRINT OF *TRAJAN'S HOLLOW 1.5* INSTALLATION

Trajan's Hollow 1.5

*The natural
deposits of gypsum
in the north of Paris
(limestone occurred
in the south)
would both offer a
material, "plaster of
Paris," and define
an ethos.*

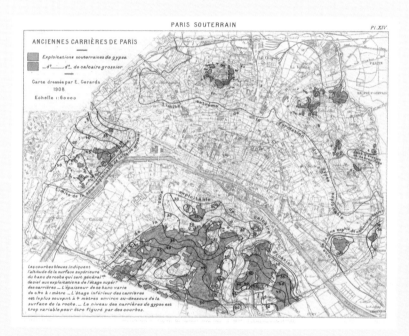

E. GERARDS, *QUARRIES OF PARIS*, 1908

PLASTER: OF PARIS TO ROME (AND BACK)

Plaster has been flowing since the middle of the third millennium BCE. The technique of plaster casting was invented by the Egyptians and replicated by the Greeks and later the Romans (most notably in the reproduction workshops of Baiae, the seaside resort on the Bay of Naples). While the production of plaster casts dates to antiquity, the collection of casts is a phenomenon that began in the 16th-century. Copies of antique statuary were collected by fine art schools so that students might study and copy or sketch the mastery of the ancients. The cast gallery (or gipsoteca) is an invention of the 19th-century that employs techniques of mechanical reproduction coupled with the commercial trading of artifacts to amass an encyclopedic and diachronic menagerie.

It is the French who first enabled the literal flow of plaster throughout the regions around Paris and then conceptually into Rome and back out again to Northern Europe and the Americas. The city of Paris was founded above a major deposit of gypsum, which, once quarried and rendered into plaster, would become known as "plaster of Paris." Over time the hills and quarries of Paris slowly transferred their gypsum deposits to the grand museums and the homes of the bourgeoisie. Plaster of Paris became the material that gave form to the ethos of the era as material extraction within the city would parallel the extraction of culture from Southern Europe.

The Académie de France à Rome, founded by Louis XIV in 1666, is the institution upon which all other academies in Rome are modeled. When Napoleon Bonaparte revived the Académie in 1803 (moving it to Villa Medici), he intended to create a base in Rome from which artists could study and reproduce the classics. These reproductions, or *envoies*, were a requirement of artists' stay at the Academy and were regularly shipped back to the capital for review. Artists were, in effect, employed as fabricators of duplications in service of the First Empire. However, the act of copying was also meant to benefit the artist. This tradition would continue on a trajectory leading to the cast galleries of the 19th-century. In this setting for artistic study, the seam lines indexing the casting process typically remain whereas they might often be removed in more public museums focused on history and culture.

The Villa Medici still maintains a gypsothèque which displays, among many other casts, a set of seventy-six pieces cast from Louis XIV's mold of Trajan's Column. The casts are displayed separately, each as their own artifact, more for a study of technique and form than narrative or historical survey.

THE GIPSOTECA OR CAST GALLERY

The cast gallery (or gipsoteca) is
an invention of the late 19th-century
that continues the tradition of return-
ing to classical antiquities for insight.
These collected reproductions of
sculptures (usually Egyptian, Greek,
Roman, or Etruscan) constitute a
material history for reference and
study. The cast gallery is often well
removed from its source material
(more often in England or the United
States than in Italy or Greece) and
offers the possibility to more closely
scrutinize the object without the
"interference" of its original context—
removed from the noise and dirt
of its environment, history can then
be properly documented.

While some cast galleries, such
as that of the Victoria and Albert
Museum in London, attempt to
reproduce "accurate" depictions
of significant statues or architectural
elements, other institutions, usually
schools of the fine arts, are more
interested in exhibiting sculptural
form than historical sampling.

*Opened in 1873,
the Cast Courts
at the V&A
constitute one
of the few
surviving examples
of 19th-century
cast galleries.*

ONE OF THE CAST COURTS AT THE VICTORIA AND ALBERT

MUSEUM IN LONDON

*The plaster casts
of the Column's
frieze are segmented
based on narrative
scenes as well
as the necessities
of geometry
and form.*

EUR (GIPSOTECA OF HISTORY)

The Museum of Roman Civilization
has become internationally known for
its historic and ongoing production
of casts of antique statuary and, in
a bizarre twist, the reproductions
housed in this museum have become
some of the most sought after
statuary in Rome. Since the originals
housed in Rome's more famous
museums (such as the Capitoline)
are now completely off-limits for use
as models, this museum's copies
have become the source from which
all further casts must be produced.
Museums from around the world
solicit copies of the reproductions
in the Museum of Roman Civilization's. Similarly, the casts of Trajan's
Column have escaped the effects of
acid rain and weathering and are now
in much better condition than the
original Column. These casts tend
to be the preferred site of study for
archaeologists and historians.

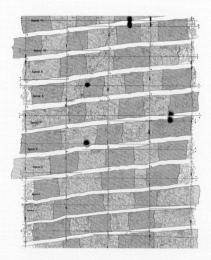

DIFFERENT CAST SEGMENTS WITHIN THE 1861 PLASTER

REPRODUCTION OF TRAJAN'S COLUMN

COPY OF THE COLOSSAL HEAD OF CONSTANTINE IN

THE MUSUEM OF ROMAN CIVILIZATION, ROME

LA SAPIENZA (GIPSOTECA OF FORM)

The gipsoteca in the School of Fine
Arts at University of La Sapienza,
Rome is exemplary of similar
collections housed in other schools
of the fine arts. These statues are
meant to be examined by students as
specimens of classical aesthetics and
expertise in the plastic arts.

Because they are not intended to
support a didactic anthropological
study of history, these casts do not
need to replicate the appearance
of the original statues. The dramatic
play of light and shadow on
the satin white plaster trumps any
desire for mimicking the historical
material effects. Seam lines from
the casting process remain and offer
more formal complexity.

*In addition
to the seam lines,
which evidence
the casting process,
the redundancy of
these sculptural
forms makes it clear
that these casts do
not pretend to be
anything other
than copies.*

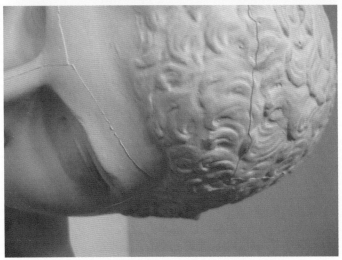

PLASTER CASTS IN THE GIPSOTECA OF THE SCHOOL OF FINE ARTS
AT THE UNIVERSITY OF LA SAPIENZA, ROME

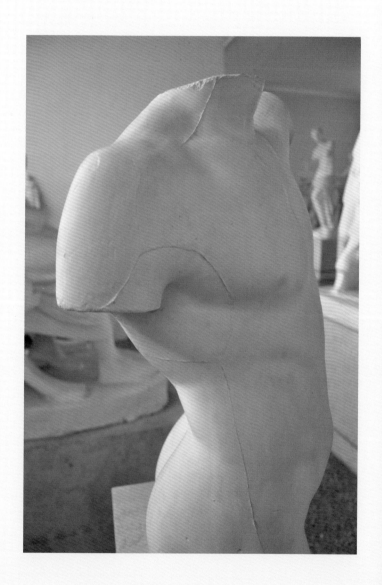

*To avoid undercuts,
a cast of the human
body can require
upwards of 200
unique mold parts,
each of which leaves
its textural residue
as seam lines.*

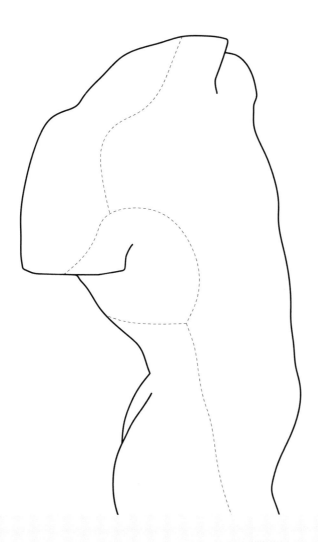

SEAM LINES REMAINING ON PLASTER CAST OF HUMAN FIGURE

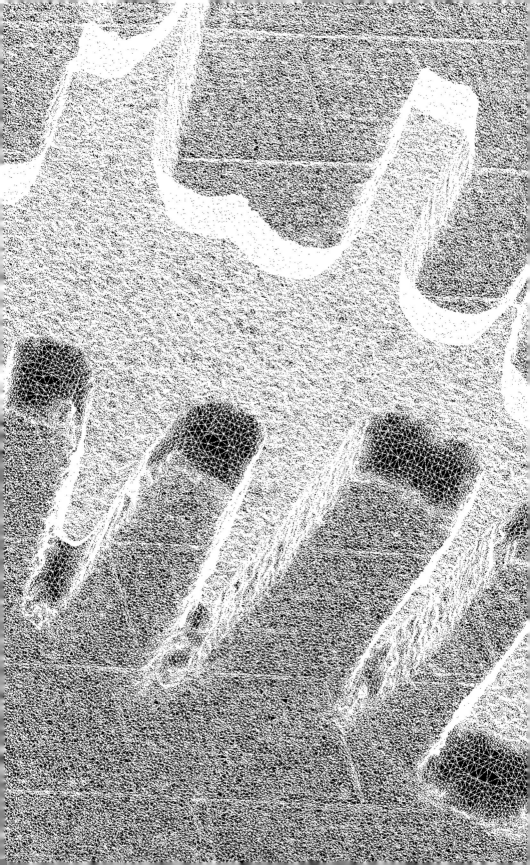

TRAJAN'S HOLLOW AS INSTALLATION

Scale = 150%

While the miniature version of the drum (*Trajan's Hollow 1:10*) tracks and updates the facture of the original Column, *Trajan's Hollow 1.5* continues the material tradition established by the series of casts and reproductions of the monument. This jump from marble to plaster implies a shift in technique and craft traditions, from stonemasonry to mold and cast systems. The strategies of subtractive carving coupled with additive stacking are replaced with the corralling and calcification of liquid into solid.

While previous plaster cast versions of the Column focus on reproducing the exterior complexity of the bas-relief narrative frieze, *Trajan's Hollow 1.5* instead attempts to capture the complex relationship between this exterior surface, the Column's massive construction, and its interior voids. Rather than a superficial index of the monument's celebrated exterior, this cast is a horizontal cross-section of the Column that reveals its least-known aspects. This prosciutto-thin slice of 5 centimeters reproduces only 1/900th of the Column's height, but contained within this micro layer are all of the most important aspects of its interior experience: the massive walls and core, the spiral stairs eroded by footsteps and cascading water, the wedge-like spaces of the window embrasure, the etched patterns of chisel marks, and of course, the apertures to the exterior surface and city beyond.

The enlargement of Trajan's Column to 1.5 times its original size allows for an amplification of the existing experience that has typically been apparent and accessible to only a handful of people. More directly, this plaster cast follows the specifics of the narrative frieze, translating the Column's ornamental surface into phenomenological depth.

This cast of the original Column fully engaged the space of the gallery, leaving similar areas of space for inhabitation both within the interior of the Column footprint and outside. The narrow wedges of space between the gallery walls and the plaster floor cast invited visitors to step within the installation and inhabit the "hollow" of Trajan's Column.

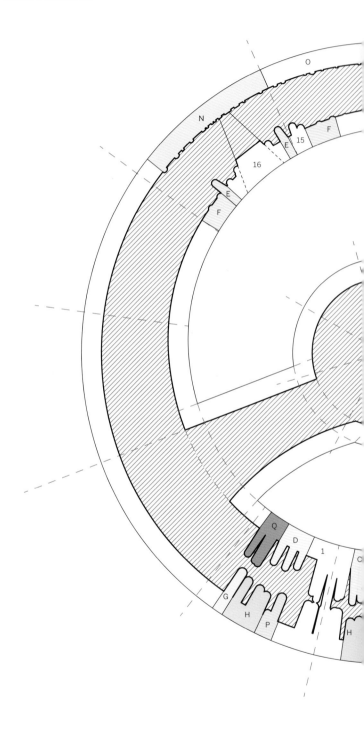

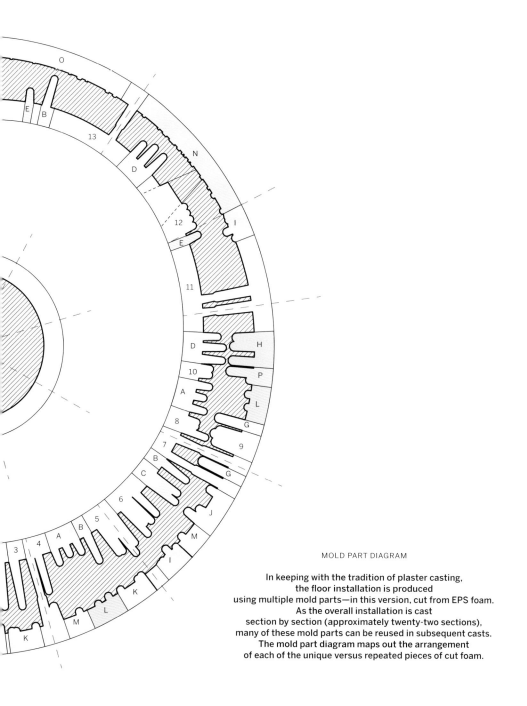

MOLD PART DIAGRAM

In keeping with the tradition of plaster casting,
the floor installation is produced
using multiple mold parts—in this version, cut from EPS foam.
As the overall installation is cast
section by section (approximately twenty-two sections),
many of these mold parts can be reused in subsequent casts.
The mold part diagram maps out the arrangement
of each of the unique versus repeated pieces of cut foam.

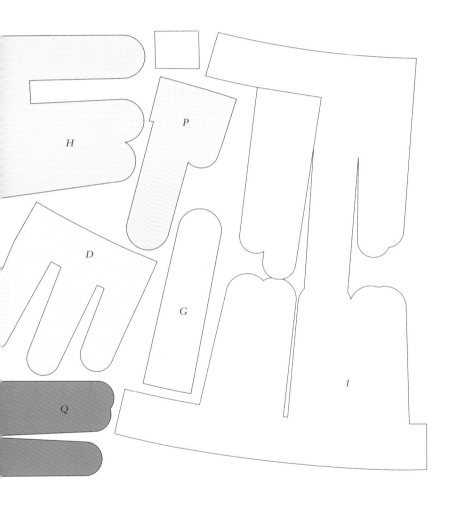

PLASTER CROSS-SECTION

The horizontal plaster bas-relief
is a section slice through the revised drum of *Trajan's Hollow 1:10*,
offering the experience of the interior space
and its new connectivity to the exterior narrative.

The multiple plaster pours fuse to form a monolithic cast, reproducing
the massing of the Column's solid core, drum wall, and stair.

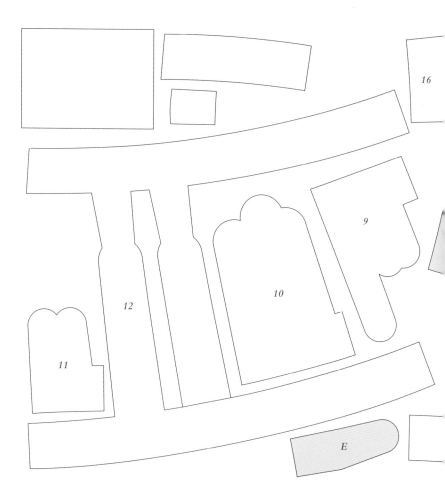

MATERIALS

500kg of plaster
Four 4' × 8' sheets of 5cm thick EPS foam
Mold release
Double-sided tape

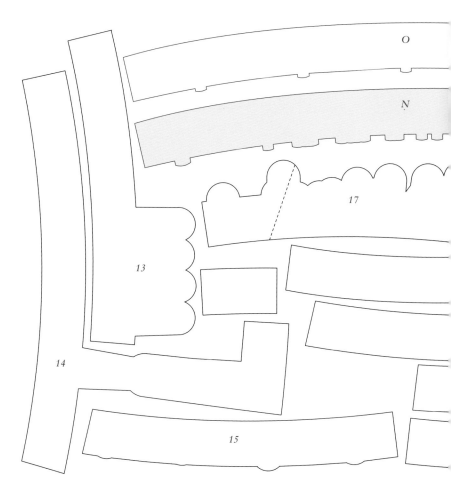

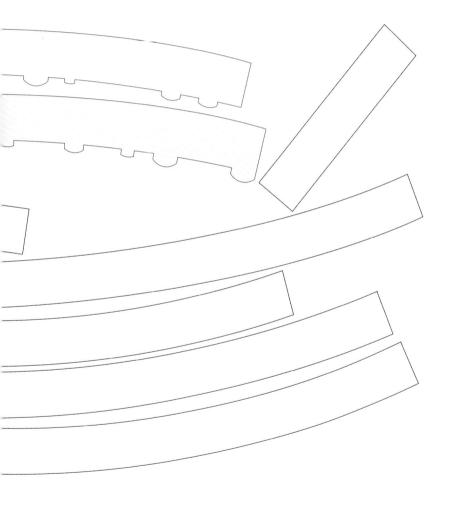

TOOLS

Hand-held hot wire foam cutter
Straightedge
Mallet
Chisel
Large plastic bucket
Wire brush
Orbital sander

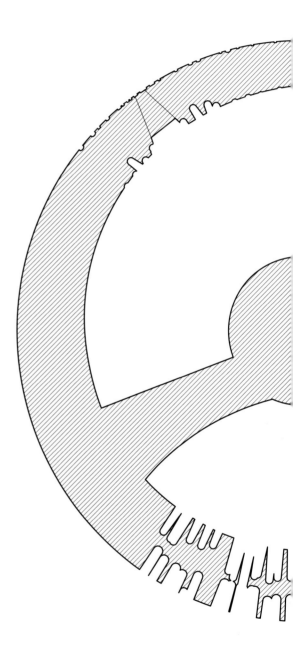

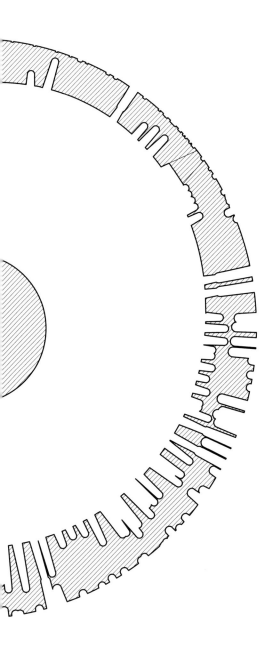

Trajan's Hollow Final Installation

RESIDUE OF PROCESS

The bas-relief floor cast of *Trajan's Hollow 1.5*
is created from multiple plaster pours
using a system of reusable foam walls and inserts.

The logic of this process remains evident in the final cast as the fins of plaster
that seeped through the mold seams have been preserved.

RESIDUE OF PROCESS

The crenellations from hand-cutting the foam mold parts update
and complicate the cuts that originate in the smaller *Trajan's Hollow 1:10*.
In addition to producing new formal information, these modifications offer
new associations to distant landscapes or the mineral origins
of the monument. They also echo the micro-landscapes of chiseled texture
in the original apertures of the Column.

THE VIRTUAL HAPTIC OF DIGITAL SCANNING

Digital scanning techniques map intricate surfaces much as a hand or liquid plaster might trace the textures and folds of ancient statuary. While these techniques capture complex surface details, they struggle to resolve coherent profiles that are so common in reproductions based on visual likeness.

This scan of Trajan's Hollow produced through photogrammetry techniques (software that computes 3D geometry by analyzing a series of 2D photographs) fuses the texture of the tiled flooring along with the plaster cast. Ruptures in the digital model are produced by view shadows or areas that lack the overlapping information necessary to "stitch" various images into three-dimensional form. Other gaps in geometry betray the perspectival locations from which photographs were taken.

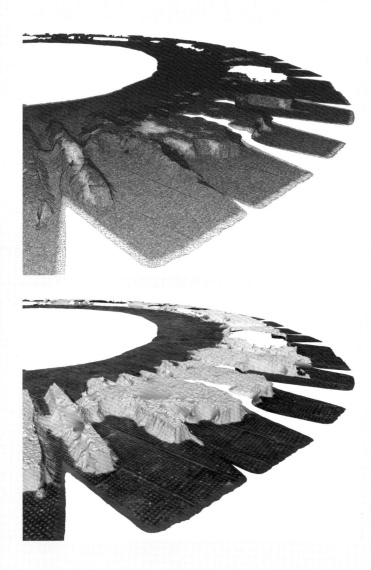

PHOTOGRAMMETRY SCAN OF *TRAJAN'S HOLLOW 1.5*
AS WIREFRAME GEOMETRY (TOP) AND IMAGE-MAPPED GEOMETRY (BOTTOM)

Endnotes

ACKNOWLEDGMENTS

This book is dedicated to Jen, my true Rome Prize. Thank you for asking me to understand history as a generative and radical force and thank you for living it with me.

This project was supported by a Woodbury University Research Grant. I appreciate the continued support of all my colleagues over the years.

Thank you to the contributors who produced original content for this book: David Gissen, Michael Swaine, Michael J. Waters, and Arcangelo Wessells. A special thanks is due to Rick Valicenti and Kyle Green at Thirst Design, without whom the book would not be properly formed, visually or conceptually.

I am indebted to those who generously read through different articles in this book: Jordan Geiger, Mark Ericson, and Mitchell De Jarnett who served as a de facto editor in the early phases of the publication. Jennifer Scappettone's help was essential at every phase of the project, from discussing overarching concepts to hashing out edits and minutiae. A host of others offered feedback or help on various aspects on the research: Marc Neveu, Michael J. Waters, Michael Swaine, and Paulette Singley who, many years ago, astutely pointed me towards Didi-Huberman's *L'Empreinte*. Thurman Grant's sharp eyes and generous input made this project feel like a welcome continuation of our first book project.

Thank you to Gordon Goff at ORO Editions and Applied Research + Design Publishing for believing in the project and to Jake Anderson for making sure it was properly executed.

Thanks to those who published aspects of the work in earlier forms: Lyn Hejinian, Christopher Patrick Miller, Glenn Adamson, and Jennifer L. Roberts.

While at the American Academy in Rome, I was fortunate enough to have both intellectual and logistical support. Thank you to the many advisors, administrators, and staff who enabled this work to thoroughly engage the city and its history. Over many dinners at the Academy and tours throughout Rome, I was able to engage many of the real experts of the Column: Mark Wilson Jones, Lynne Lancaster, and Peter Rockwell, as well as the expert on scanning antiquity, Bernie Frischer. Thank you for generously sharing with me your many years of research.

Thanks to those who helped with documentation during the expedition up the Column: Seth Bernard, Andrew Riggsby, and Michael J. Waters and to Giovanna Latis and Michael J. Waters for generously spending hours photographing the installation. Thanks to Mitchell Bring for his photogrammetry expertise.

Thanks to all those in the AAR community who enthusiastically volunteered to help cast plaster (or deconstruct it), especially Emily Morash, Michael J. Waters, Paul Rudy, and Adrian Van Allen. Others wandered in regularly to offer advice and engaging conversations. Thank you to William Kentridge for the daily challenges to continually cast more plaster.

And to Harry, Kristine, Hilary, and Florentina for always asking the right questions with the right amount of enthusiasm.

IMAGE CREDITS

All images and drawings by Joshua G. Stein / Radical Craft unless otherwise credited

cover: design by Thirst, image by Jason Kwong

p. 2: Michael J. Waters

Foreword

p. 6: Giovanni Battista Piranesi

Introduction

p. 12: all images by Michael J. Waters

Part I: Monuments

Trajan's Column

pp. 27–35: All images by Giovanni Battista Piranesi from *Trofeo o sia Magnifica Colonna Coclide (The Trophy or Magnificent Spiral Column)*

"On Architectural Materiality" **p. 39:** drawing by Michael J. Waters (after Mike Bishop) **p. 42:** Michael J. Waters **p. 43:** drawing by Michael J. Waters (after Lynne C. Lancaster) **pp. 46–47:** drawings by Michael J. Waters (after Silvana Rizzo)

"Pocket Landscapes" **pp. 48, 50 (bottom), 52, 55:** Michael J. Waters **p. 50 (top):** Giambattista Nolli **p. 58:** Jason Kwong

Hidden Trajan

p. 64: engraving by Giovanni Battista Piranesi modified by Joshua G. Stein **pp. 65–67:** All images by Seth Bernard **p. 69:** image by Giovanni Battista Piranesi modified by Radical Craft **pp. 70, 72–73:** Michael J. Waters **pp. 75–77:** drawings by Peter Rockwell, images by Andrew Riggsby, matrix by Radical Craft **pp. 79–117:** all images by Michael J. Waters, diagrams by Radical Craft

Trajan's Progeny

p. 122: photo composite by Kyle Green **pp. 120, 124 (top):** "Column. of.marcus.aurelius.complete.arp," by Adrian Pingstone (Arpingstone), cropped from original image **pp. 120, 124 (middle):** "Karlskirche - Wien 006," by DALIBRI, [1] cropped from original image **pp. 120, 124 (bottom):** "Paris Colonne Vendôme 2012 1," by Lionel Allorge, [1] cropped from original image **pp. 120, 125 (top):** "Converted," by ScottyBoy900Q, [1] cropped from original image **pp. 120, 125 (middle):** "Congress Column at Christmas," by www.flickr.com/photos/infoma-tique/, [2] cropped from original image **pp. 120, 125 (bottom):** George A. Grant, cropped from original image **pp. 121, 128 (bottom):** © Victoria and Albert Museum, London **pp. 121, 129 (top):** "TrajanscolumbMdCR," by Notafly, [1] cropped from original image **pp. 121, 129 (bottom):** Michael J. Waters

"The Wayward Cast"
p. 130: Francesco Carradori
p. 135: All images © Victoria and Albert Museum, London **p. 136:** "This_room_is_full_of_plaster_cast_copies_of_famous_world_statues,_sculptures_and_monuments," by M.chohan, public domain **p. 141:** Giovanna Latis **p. 142:** "San Carlo alle Quattro Fontane (Lugano)" by Little-Joe, [1] cropped from original image **p. 145:** All images © FAT (Fashion Architecture Taste)

Part II: Reconstructions

"A Set of Directions for the Reader"
pp. 151–159: All images by Arcangelo Wessells

Trajan's Hollow 1:10

"Marble in the 21st-Century"
p. 162: Domenico Zaccagna **p. 167:** John Bartholomew

p. 177 (top): drawing by Peter Rockwell **pp. 182–195:** All images by Jason Kwong

Scaling Trajan's Hollow

"The Agency of Scale"
p. 198: Achille Collas **p. 200 (top):** "Pointing machine," by Satrughna, [1] cropped from original image **p. 200 (bottom):** Diderot et d'Alembert **p. 201 (top):** photo courtesy of the Iris Cantor Collection **p. 201 (bottom):** "Rodin 03," by Mgmoscatello, [1] cropped from original image **p. 215 (left):** © Claes Oldenburg, photo by David Heald © Solomon R. Guggenheim Foundation, NY, courtesy of the Oldenburg van Bruggen Studio, cropped from original **p. 215 (right):** © Claes Oldenburg, photo courtesy of the Oldenburg van Bruggen Studio

Trajan's Hollow 1.5

"Plaster: of Paris to Rome (and Back)"
p. 226: E. Gerards **p. 229:** "Another Room of Casts," by M.chohan, public domain **p. 230:** Radical Craft over drawing by Peter Rockwell **p. 231:** "Museo della Civiltà Romana – Sala di Costantino," by sono io l'autore della foto, public domain

pp. 262–263, 274–277: all images by Michael J. Waters **pp. 268, 269:** all images by Walker Smith-Williams

Index

pp. 293, 294–295: engraving by Giovanni Battista Piranesi

1 used under CC BY-SA 3.0 (http://creativecommons.org/licenses/by-sa/3.0)

2 used under CC BY-SA 2.0 (https://creativecommons.org/licenses/by-sa/2.0/)

BIOGRAPHIES

Joshua G. Stein

Joshua G. Stein is the founder Radical Craft and the co-director of the Data Clay Network (www.data-clay. org), a forum for the exploration of digital techniques applied to ceramic materials. Radical Craft (www.radical-craft.com) is a Los Angeles-based studio that advances design tactics steeped in history—from archaeology to craft—to produce contemporary urban spaces and artifacts while evolving newly grounded approaches to the challenges posed by virtuality, velocity, and globalization.

Stein is co-editor of *Dingbat 2.0*, the first full-length publication on the iconic Los Angeles apartment building type and has received numerous grants, awards, and fellowships, including multiple grants from the Graham Foundation for Advanced Studies in the Fine Arts, the AIA Upjohn research award, and the 2010–11 Marion O. and Maximilian E. Hoffman Rome Prize Fellowship in Architecture. He is a former member of the LA Forum Board of Directors and has taught at the California College of the Arts, Cornell University, SCI-Arc, and the Milwaukee Institute of Art & Design. He is Professor of Architecture at Woodbury University where he also directs The Institute of Material Ecologies (T-IME).

David Gissen

David Gissen is the author of books, essays, exhibitions, and experimental writings and projects about environments, landscapes, cities, and buildings from our time and the historical past. Gissen is a professor at the California College of the Arts (CCA) and former visiting professor at Columbia University's Graduate School of Architecture, Planning and Preservation and the Massachusetts Institute of Technology, History, Theory, Criticism program. He lectures and teaches in the areas of architecture, urban and landscape history-theory, preservation, writing, and design. At CCA, he co-directs the Experimental History Project and the MAAD History, Theory, Experiments degree.

Michael J. Waters

Michael J. Waters is an Assistant Professor in the Department of Art History and Archaeology at Columbia University. He earned his PhD from the Institute of Fine Arts, New York University, and was previously the Scott Opler Research Fellow at Worcester College, University of Oxford. He has also held fellowships at the Villa I Tatti, American Academy in Rome, Scuola Normale Superiore di Pisa, and Sir John Soane's Museum.

His forthcoming book, *Renaissance Architecture in the Making*, examines how materials, methods of facture, building technology, and practices of reuse shaped the development of fifteenth-century Italian architecture. He has also published articles on the study of antiquity and early modern architectural prints, drawings, and treatises, and in 2011, he co-curated the exhibit "Variety, Archeology, and Ornament: Renaissance Architectural Prints from Column to Cornice," at the University of Virginia Art Museum.

Michael Swaine

Michael Swaine is an artist working in a variety of materials, methods, and media with a long-time focus on collaborative work—in particular with Futurefarmers. Swaine has participated in exhibitions at the Cooper-Hewitt, National Design Museum, New York; San Francisco Museum of Modern Art, California; SITE Santa Fe, NewMexico; The Guggenheim, New York; among others. His Free Mending Library received notice and a Certificate of Honor from the Mayor of San Francisco.

Swaine holds a BFA from Alfred University and an MA from the College of Environmental Design at University of California, Berkeley. He most recently taught at California College of the Arts and Mills College. Swaine began teaching at the University of Washington in 2015.

Rome Design Team

Joshua G. Stein: Lead Design
Walker Smith-Williams: Project Design and Management
Federico Giacomarra: Parametric Modeling
Luca Petricca: Parametric Modeling
Yujin Oh: Installation
Jarupa (Orm) Tachasirinugune: Installation
Kristie Huey: Graphics

Publication Assistance

Rebecca Fox
Jessica Gardner
Axel Olson

For more information on the work of Joshua G. Stein and Radical Craft, visit www.radical-craft.com

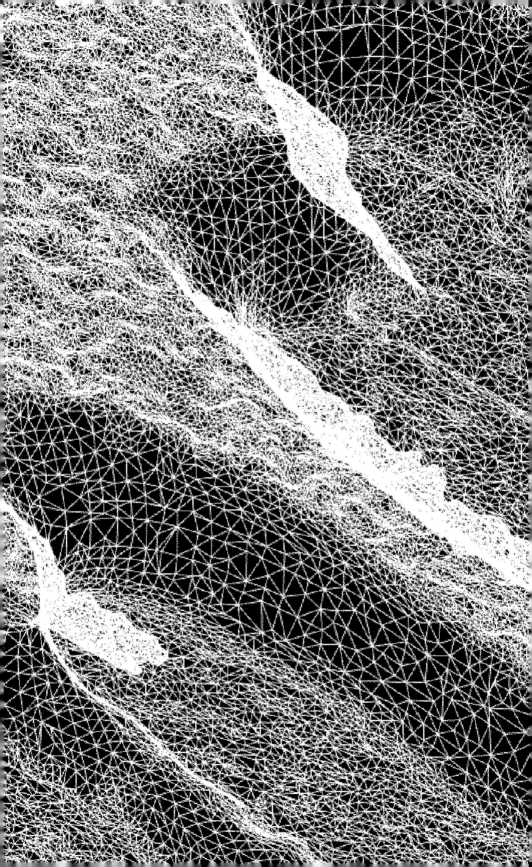

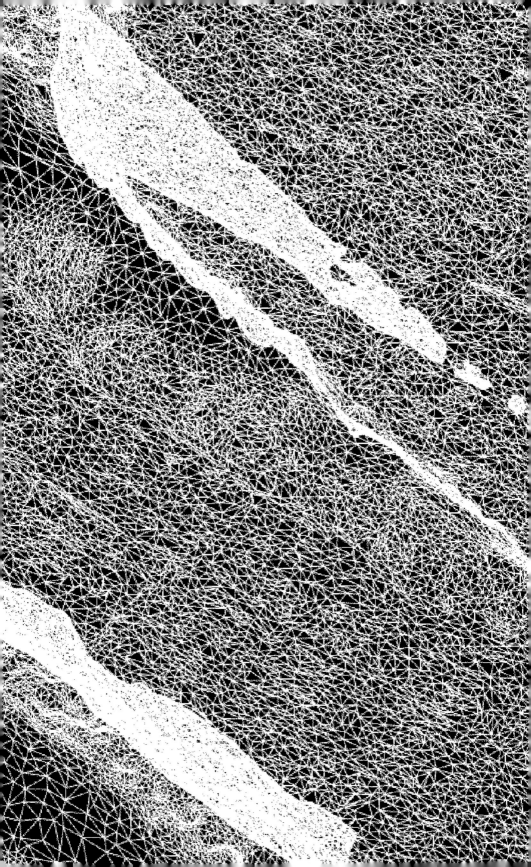

INDEX

"A FACE OF THE PEDESTAL OF TRAJAN'S COLUMN", PLATE XI

FROM *THE TROPHY OR MAGNIFICENT SPIRAL COLUMN* (1774–75), G.B. PIRANESI

(DETAIL ON FOLLOWING SPREAD)

Published by Applied Research and Design Publishing. An Imprint of ORO Editions

Gordon Goff: Publisher
appliedresearchanddesign.com

Text by Joshua G. Stein, David Gissen, Michael Swaine, and Michael J. Waters.

Design by Kyle Green and Rick Valicenti at Thirst. Typefaces used include, Bembo designed by Monotype and Benton Sans designed by Font Bureau.

Project Manager: Jake Anderson

10 9 8 7 6 5 4 3 2 1 First Edition

ISBN: 978-1-940743-93-6

Color Separations and Printing: ORO Group Ltd.

Printed in China.

AR+D Publishing makes a continuous effort to minimize the overall carbon footprint of its publications. As part of this goal, AR+D, in association with Global ReLeaf, arranges to plant trees to replace those used in the manufacturing of the paper produced for its books. Global ReLeaf is an international campaign run by American Forests, one of the world's oldest nonprofit conservation organizations. Global ReLeaf is American Forests' education and action program that helps individuals, organizations, agencies, and corporations improve the local and global environment by planting and caring for trees.